Contents

Introduction to SuperCollider

Introduction to
SuperCollider

Andrea Valle

Introduction to SuperCollider

Bibliographic information published by the Deutsche Nationalbibliothek

The Deutsche Nationalbibliothek lists this publication in the Deutsche Nationalbibliografie; detailed bibliographic data are available on the Internet at http://dnb.d-nb.de.

ISBN 978-3-8325-4017-3

Logos Verlag Berlin GmbH
Comeniushof, Gubener Str. 47,
10243 Berlin
Tel.: +49 (0)30 42 85 10 90
Fax: +49 (0)30 42 85 10 92
Web: http://www.logos-verlag.de

PDF version:
site: http://logos-verlag.de/ebooks/IntroSC/
user: introsc
password: Super0616Collider!

Translated from Italian by Marinos Koutsomichalis and Andrea Valle.
English revision by Joe Higham and Joshua Parmenter.
Italian edition:
Apogeo Education, Maggioli editore, Milano, 2015
ISBN: 978-88-916-1068-3

Typeset with ConTeXt and SuperCollider by Andrea Valle

Foreword

My first encounter with SuperCollider dates back to 2002, thanks to my friend Hairy Vogel, one of its first (and bold) users. At that time, I remember being deeply impressed by the sound quality, the seamless integration between audio synthesis and algorithmic composition, the native real-time operating mode. So, I decided to buy an Apple computer, as SuperCollider 2 was working on MacOS9 systems only, while SuperCollider 3, that had just become open source, was still an unstable release, porting to linux had just begun (as far as I can remember) and there were no plans at all for a Windows version.. But in 2002 I was lacking some basic skills to be productive on SuperCollider, so I left it momentarily, to get back to it in 2005. In 2008 I switched to SuperCollider in my class for the Multimedia and Arts program (DAMS) at the University of Turin. I was not able to find a basic, yet comprehensive resource to use for teaching. Comprehensive of what? This is of course the main point. My class is a basic introduction to computer music, so I was in need of reference for both some basic notions in DSP *and* SuperCollider. Still, I was confident that the latter was the apt tool to approach the former. Yet, introducing SC is to also introduce computer languages in general and certain notions concerning computer science *tout court* (e.g. the client/network architecture). That is why I decided to write this book the scope of which is, consequently, too broad to offer completeness in any of each subjects. From a very personal perspective, I have to confess that I have tried to write the book that I wished to have while starting with SuperCollider. My hope is that someone might be in need of it as I was.

Originally written in Italian in 2008 in an electronic only version, the book has been substantially re-edited and updated to account for the SC 3.7 version in 2015, and then, revised and translated into English in 2016 to be published in a printed version thanks to Logos Verlag. It has been translated into English by myself and Marinos Koutsomichalis, thanks also to the support of the SuperCollider community. A crucial help on language and content revision has been provided by Joe Higham and Joshua Parmenter. I am deeply thankful to Marinos, Joe and Josh for their work.

SuperCollider has a strong, supporting and passionate community. Without it, my journey with SC probably would have ended very soon. So, I am profoundly indebted to all its members. Finally, a special "grazie" to James McCartney to have his ideas supercollide in his work.

Ciriè, Turin, May 25, 2016

1 Getting started with SuperCollider

1.1 About SuperCollider

SuperCollider (SC) is a software environment for real-time audio synthesis and interactive control. It represents the state of the art in audio programming technologies: as of writing, there seems to be no other software solution which is, at the same time, equally effective, efficient and flexible. Yet, SC is often, and rightly, approached with suspicion and with awe: SC is not exactly "intuitive". In the first place, the interface is textual. In SuperCollider the user is supposed to write, which is difficult enough for the typical computer-user who is normally used to seeing (and interacting with GUI) and not in manipulating and reading raw code. Then, like all software solutions to advanced audio computing, SuperCollider prerequisites in sound synthesis and algorithmic music composition, as well as in computer programming. More, it demands thorough understanding of certain informatics-derived concepts. However, these ostensible obstacles are in reality the true power of SC: there would be no other way to arrive at a system equally versatile in software design, efficient in computer modeling, and expressive in terms of easy to use symbolic representations.

SuperCollider was originally developed by James McCartney in 1996 for the Macintosh platform. Version 2, in particular, has been designed for the MacOS 9 operating system. SuperCollider 3 (which is, at the same time, very similar and yet very different to SC2) has been developed initially for the MacOS X operating system, but is now an open source project that revolves around a community of contributing programmers. This community of developers has been responsible for porting SC in Windows and Linux operating systems. Accordingly, we witnessed major changes in the software during the last years, not in terms of the core functionality (which albeit having been improved, is effectively the same) but more specifically the user interface. The first versions of SC for other platforms have been less featured than the OSX version. The various Linux versions have been gradually aligned with the OSX version in terms of functionality, yet and until recently, their user interface has been very different. On this aspect, the various Windows versions used to be less intuitive than the Linux/OSX versions in various respects. This situation has been radically changed with the introduction of the new Integrated Development Environment (IDE) (as from version 3.6) and of the new help-system (as of version 3.5). From 3.6 onwards, SC looks and works identically on all three platforms, this way providing a consistent user experience[1].

What can we do with SuperCollider? The SC community is highly diverse, SC being successfully used for live improvisation, GUI-design, dance-oriented music, electroacoustic music composition, sound spatialization, audio-hardware interaction, scientific sonification, multimedia or network based applications, the performance of distributed systems and for all sorts of other applications. There have been real-life examples on all the aforementioned cases that have been made possible either with SC alone or with respect to paradigms that emerged from within it. For instance, the so-called "live coding" paradigm—an improvisation practice where the instrumental performance (to use a contextually uncanny, albeit not erroneous expression) is based on the writing/editing of code in real-time to control sound-generating processes[2]. Live coding practices that are now being encountered in various other environments, have emerged within SuperCollider (they are, somehow, inherent to its nature). Another "esoteric" and SC-specific project is SC-Tweet: musicians are supposed to compose using code of a maximum size of 140 characters (the size of a "legal" tweet). It has to be noted that such microprograms often result in surprising complex

---

[1] In reality, SC is highly customizable with respect to intrinsic language features.
[2] The reference site is
    http://toplap.org
    .

sounds/textures[3]. Therefore, there are many good reasons to learn how to use SuperCollider.

1.2 SC overview

In a nutshell, SuperCollider is a software environment for the synthesis and control of audio in real time. However, "Software environment" is a rather vague term. The official definition (the one that used to appear on the official website's homepage) is more precise/comprehensive:

> "SuperCollider is an environment and programming language for real time audio synthesis and algorithmic composition. It provides an interpreted object-oriented language which functions as a network client to a state of the art, real-time sound synthesis server."

In detail:

1. *an environment*: SC is an application comprised of individual and disparate components. Hence the use of the term "environment".

2. *and*: SC is also something else, something completely different.

3. *a programming language*: SC is also a programming language. As to be further discussed, it belongs to the broader family of "object-oriented" languages. In order for SC code to be operational, it has to be interpreted by a special software module called an "interpreter". An interpreter is a program that "understands" some programming language and that commands the computer to act accordingly. SC is also a language interpreter for the SC programming language.

---

[3] The first occurrence of SC-Tweet is for the *sc140* album published in 2009 by the The Wire music magazine*The Wire*

http://www.thewire.co.uk/audio/tracks/supercollider-140.1?SELECT\%20*\%20\%20FROM
\%20request_redirect

.

4. *for real-time audio synthesis*: SC is optimized for the synthesis of real-time audio signals. This makes it ideal for use in live performance, as well as, in sound installation/event contexts. It is still possible to use SC in order to generate audio in non real-time, but such a use is less immediate and, in practice, it is encountered less often.

5. *and algorithmic composition*: one of the strengths of SC is to combine two, at the same time both complimentary and antagonistic, approaches to audio synthesis. On one hand, it makes it possible to carry out low-level signal processing operations. On the other hand, it does enable the composers to express themselves at a much higher level; that is, not in terms of audio samples, but rather in terms of higher level abstractions that are more relevant to the composition of music (e.g.: scales, rhythmical patterns, etc). In that sense, SC is ideal for purely algorithmic or formal approaches to composition. In SC, that kind of operations can be performed interactively and in real-time.

6. *[the] language [...] functions as a network client to a [...] server*: SC interpreter (the application that interprets the SC language) is also a client that communicates through a network to a server—a service provider.

7. *a state of the art*: SC currently represents the state of the art in audio programming: there is no other software package available that is equally powerful, efficient, or flexible (and as of 3.6, also portable).

8. *sound synthesis server*: SC is also a provider of services, in particular of audio synthesis services. The phrase may seem mysterious: SuperCollider may generate real-time audio on demand. Those software processes or human individuals that request real-time audio are commonly referred to as "clients".

   In summary, SC may (confusingly enough) refer to six different things, as outlined in Figure **1.1**:

1. an audio server (a provider of real-time audio on-demand)
2. a programming language
3. an interpreter (a program that interprets) for the above language
4. a client for the audio server

5. an application bundle comprised of all the above (1-4) components
6. and that also includes a special Integrated Development Environment (IDE), that is, an editor where code may be written/edited and which forwards code to the interpreter for evaluation.

Accordingly, SC application consists of three parts: the audio server (referred to as scsynth); the language interpreter (referred to as sclang) and which also acts as a client to scsynth; and the IDE. Installing SC is to install this entire scheme—these three applications are bundled together and communicate internally with each other when needed[4]. The IDE is built-in this application bundle. The server and the client (language interpreter) are two completely autonomous programs—this will be elaborated upon later: for now, keep in mind that these are two separate programs, and that when you install SC you get two programs at the price of 1 (the actual cost is calculated as follows: $2 \times 0 = 0$: as a madrigal by Cipriano de Rore says, "my benign fortune").

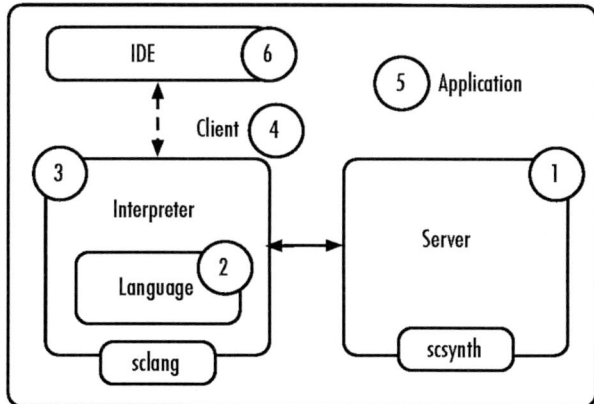

Figure 1.1   Structure of SC.

This may seem complicated and, indeed it is.

1.3 Installation and use

The official SC site is:

    http://supercollider.github.io/

Managed by the community, it basically includes everything needed to directly use SC. It additionally features a series of learning resources (section "learning"), and several code examples (see in particular the "video" section on the site). Regarding the community, although there is significant presence in various forums and social networks, the historical mailing list is worth mentioning, since this is where we may easily encounter and/or come in contact with most of the best SC veterans:

    http://www.beast.bham.ac.uk/research/sc_mailing_lists.shtml

This is a high-traffic mailing list and immediate answers are to be, typically, expected. The list's archives are also a valuable resource; more often than not a answer to some question is already available therein.

Finally, the original site by James McCartney, which has an eminent historical value, is still online:

    http://www.audiosynth.com/

SC can be downloaded from Github, in particular from the following address:

    http://supercollider.github.io/download.html

Installation is easy when using the precompiled binary files that are available for OSX and/or Windows. For those platforms installation is similar to everyday programs and downloads and, therefore, there is no need to further examine it. Installation on Linux also follows the norms of the penguin, with slight variations according to the idioms of the chosen distribution.

In addition to the SuperCollider bundle, there are several third-party extensions available in the form of "Quarks", that may be installed from within the program itself.

When first executed, SC looks a lot like what we see in Figure **1.2**. Note that this is not the usual arrangement of graphic containers, even if the elements are the same. This already demonstrates that the IDE can be personalized as needed. On the right side there is a (empty) text field where code may be edited. On the left side, two components may be identified: the post window, where the interpreter posts the results of the various evaluated blocks of code (in this case, information related to the initialization of SC itself), and the help browser (here hidden by the post window), where one may navigate the available Help files. The right side of the bottom bar, displays information about sclang (Interpreter, active, in green) and scsynth (Server, inactive, white).

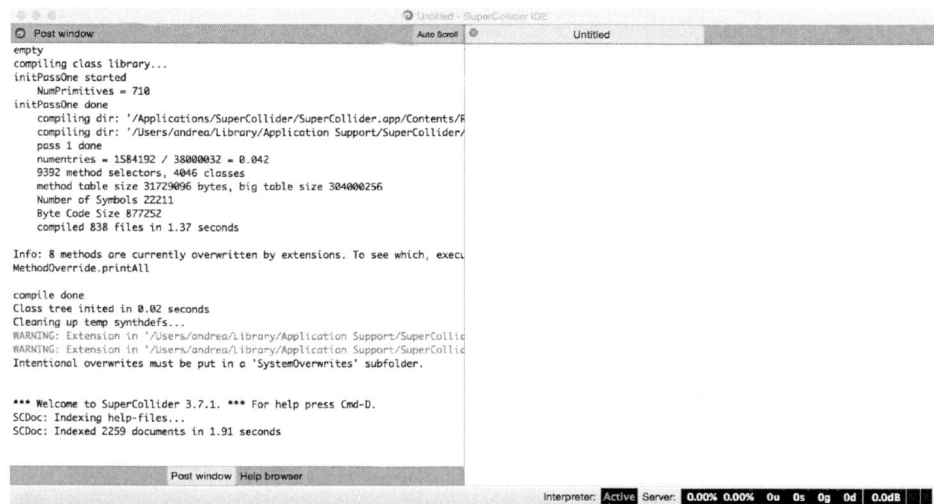

Figure 1.2 SC IDE.

IDE's menu is not particularly complex. Some quick instructions/directions, most of which will be also encountered later on, follow.

- *File* is dedicated to opening/saving SC documents. These are files in ASCII text format and can be of three possible types, depending on the extension:

  — scd is dedicated to SuperCollider Documents;
  — sc is dedicated to files with SC class definitions (to be discussed later);

— schelp is reserved for the Help files, which should be written in a special format for the Help browser to read them properly;

- *Open startup file* command opens for editing a configuration file that is invoked every time SC is launched.

- *Session*: enables the management in unitary groups files belonging to the same project.

- *Edit*: contains the usual set of entries related to the editing of text. The last section of the menu includes code-specific commands. Beyond editing commands, particularly useful is "Select Region" command.

- *View*: allows you to manage/reorganize the visual elements of the IDE (e.g. the availability of the post window, of the help browser and/or of the code window. Figure **1.2** shows a custom-configured session).

- *Language*: includes a series of SC-specific commands, divided in five blocks:

  — the first comprises Interpreter-specific commands. E.g., sclang (which, as already discussed, is a stand-alone program) can be closed or restarted from this menu;
  — the second concerns the management of the audio server (as already discussed, scsynth is also a standalone program);
  — the third features a series of commands that activate some graphical interfaces for displaying information on the audio server;
  — the fourth is devoted to the evaluation of the code, that is, to indicate what code (or portion) should be taken into account for interpretation (or for the actions that follow) ;
  — finally, the fifth includes a set of commands to interactively explore subject code SC is largely in SC, thus allowing introspection—that is to allow SC to explore its own structure.

- *Help*: allows you to interactively open Help files associated with the relevant selected elements of SC.

In summary, how to use SC: writing/editing code, selecting a line or a block of code, asking the interpreter to evaluate this code, navigating the documentation files, and interactively repeating all the previous actions during a working section. Yet, alternative and more sophisticated, working paradigms are also possible; e.g. relying on programs that themselves trigger sound processes of indeterminate duration, or to design standalone graphical user interfaces to be used interactively in performance contexts. In reality, there are several possible and disparate styles of work. The only action that cannot be omitted is to evaluate at least one block of code (which represents a single SC program) once.

1.4 Objectives, references, typographical conventions

The objective of this book is, then, twofold:

1. to provide a brief overview/introduction to SuperCollider, because as of present there are no other available resources (as far as the author knows).

2. to introduce readers to some key aspects of what is commonly, albeit inaccurately and often inappropriately, referred to as computer music.

The departing hypothesis is that both objectives are mutually relevant, and that, eventually, SuperCollider is the perfect tool to achieve the second. Since SuperCollider presupposes programming skills, this introduction is also an informal introduction to computer programming, in particular to audio/music programming via the use of the SuperCollider language.

The material presented herein is, with a few notable exceptions, "original". In particular, those parts relating to sound synthesis are based upon excerpts from the Italian handbook *Audio e multimedia*[5], an advanced but not very technical introduction to the subject. Also being an informal introduction to sound synthesis, the present, up to a certain extent, draws upon material/conveniences found in several other resources, some of which are more complex and more complete[6]. As far as the SuperCollider-specific parts are concerned, this manual is neither a translation nor an extension to any other existing text. More, it should not be understood as a substitute or alternative for the reference

---

[5] V. Lombardo, and A. Valle, *Audio e multimedia*, Milano, Apogeo 2014 (4ed.), hence on AeM

[6] This is not the place for a bibliography on computer music, in this regard please note the references in AeM, chap. 5.

text, i.e. The SuperCollider Book[7]. The latter, modelled analogously to Super-Collider's predecessor language's main manual (*The Csound Book*), comprises several chapters including tutorials, in-depth analyses and descriptions of specific applications/projects. Yet, it is not to be thought of as a unitary text to be read sequentially, which is, instead, the purpose of this book. Finally, it should be stated explicitly that the material presented herein boldly draws upon a series of resources, most importantly upon the SC help files, upon tutorials by James McCartney, Mark Polishook, Scott Wilson[8] and by David Cottle[9], and upon material derived from the invaluable SuperCollider mailing list[10]. The reading of the aforementioned resources are not to be thought of as superfluous, since the present book has a preparatory and complimentary role in respect to them.

Ordering the available materials in the case of SC is always very difficult, precisely because to some extent all (planning, architecture, audio) have to be introduced together. Since the first (Italian) edition (2008) of this book it occurred to me several times to change the chapter sequence so that sound synthesis comes first (this is typically the main reason someone is interested in SC in the first place). However, from my teaching experience, it is very risky to deliberately leave certain programmatic aspects vague until later. Instead, I herein follow a more concise approach, discussing audio synthesis only after having examined the SC language in detail.

I have tried to make the text as readable as possible, also in purely visual terms, by clearly marking the various blocks of text in different ways. Thus, the following typographic conventions are followed in the book:

1. text: character in plain black, like written here;

---

[7] S. Wilson, D. Cottle, N. Collins, *The SuperCollider Book*, Cambridge, MIT Press, 2011. Another recent addition is M. Koutsomichalis, *Mapping and Visualisation in SuperCollider*, which is a step by step discussion of very sophisticated subjects regarding SuperCollider in general, as well as its multimedia applications in particular.

[8] These tutorials are included in the standard SC distribution.

[9] *Computer Music with examples in SuperCollider 3*,
     http://www.mat.ucsb.edu/275/CottleSC3.pdf

[10] It should be also noted that the section "docs" on Github also contain several significant learning resources.

2. code: written with monospaced characters, following syntax-colorization[11], and enumerating lines. In the pdf version of this book, each example is followed by an interactive marker which will directly open the corresponding source file with the SC IDE when clicked. Readers interested in this feature are advised to use Adobe Acrobat Reader (which is a free and multi-platform).

```
1 // example
2 "something".postln ;
```

3. post window: written in black, inside an orange text box and enumerating lines[12].

```
1 an example
```

---

[11] Color remarks apply to pdf version, in printed version gray scale is in use.
[12] In some rare cases, the same format is used for non-SC related text. Such cases are to be distinguished by their context—note that the interactive marker is included in those cases, too.

2 Programming in SC

2.1 Programming languages

A computer is a device capable of manipulating symbolic entities representable in binary form. In other words, the information in a computer is represented by sequences of quantities that may have only two possible values, 0/1, on/off, and so on. The theoretical foundation of computer science is typically attributed to Alan Turing, who first proposed a formalization of an abstract machine designed as a device to read from and write into a memory. With a few notable exceptions (that have been much discussed but not really empirically tried out yet, at least not in a large scale), all contemporary computers that are broadly in use implement Turing's abstract machine by means of a technological framework that is called, by its designer, Von Neumann architecture. Once the computer is built, the problem lies in its control, that is, in defining the reading/writing operations that this machine should perform. Such operations can be performed in close contact with the machine, i.e. by directly defining the ways in which it should manage the manipulation of the memory cells. Let us consider the case of the binary representation, which is the way that all the information is described in a computer. The description of an alphabetic character –which can be written in a text editor such as that of the SC IDE– requires 7 bits (also called ASCII encoding). In other words, each character is associated with a sequence of seven 0/1. For example, in ASCII the 7-bit binary code 1100001 represents the character a. It is quite obvious that describing explicitly the information at this level (bit by bit), requires such an effort, that the enthusiasm for using a computer quickly vanishes. Instead, one of the fascinating aspects of the theory of programming lies in the fact that certain representations can be described at a higher level in a more "meaningful", "intuitive" form. You will notice that for example a is significantly more compact and intuitive than 1100001. Similarly, the character sequence SuperCollider is consequently represented by this binary sequence:

```
01010011011101010111000001100101011100100010000110110
11110110110001101100011010010110010001100101011100010
```

The alphabetical sequence SuperCollider is therefore more "abstract" and "higher level" than the binary one. If we assume that the latter represents the lowest level – that is, the only one that can be actually used to "talk" to the machine – then if we want to control the machine through the alphabetical sequence it becomes necessary to translate it into a binary sequence. How do we perform this job? It is the job of a specific "translator" program. For example, the binary representation above is obtainable in SuperCollider through the following program, which is therefore a translator for the (quite minimal) SuperCollider program into the binary program:

```
1  "SuperCollider".ascii.collect{|i| i.asBinaryDigits}
2  .flat.asString.replace(", ", "")
```

There are two consequences stemming from the previous short discussion:

The first is that when there is a programming language, there is always the need for a translator program, which is typically called the "compiler" or "interpreter".

A compiler outputs an executable "program" that can be run. For example, the SuperCollider *application*, as virtually all the programs that reside on a computer, results from the compilation of a code written in C++. In other words, the C++ code is compiled by the compiler, and the latter outputs the executable (i.e. the SC application to be launched by the user). Developers (not users!) of SuperCollider write C++ code that is used to build the overall program, which includes the core functionality for the audio, the interface with the operating system, the use of already available libraries to manage GUI elements, and so on. These codes are then compiled (they are called "sources"[1], by means of a compiler that is specific for each of the supported platforms and architectures (e.g. respectively OSX, Linux, Windows, and Intel, PowerPC, and so on). The process ends by outputting the executable (the one shown in Figure **1.1**). As SC is an open source project, this source code is available to everyone, and everyone can eventually compile it on their own computer (provided that there is a compiler installed on their machine).

An interpreter does not generate an intermediate representation which will be executed later, rather it translates and executes specified instructions directly. Because the SuperCollider *language* is interpreted, the SC code from the previous example can be evaluated in the IDE (which passes it to the interpreter) in order to be immediately executed. For the code to run in SuperCollider it must be selected, then the command "Evaluate Selection or Line" from the *Language* menu must be launched (to make things easier the user should familiarize themselves with the key combinations for shortcuts). As a consequence of the interpretation, the binary sequence above will be printed in the post window.

Therefore, the SuperCollider language is an interpreted language, and the user is thus able to interact with the interpreter. The interpretation takes place immediately (the code is executed as quickly as possible), and the interpreter remains available for the user. So, programming in SC is an interactive process, in which the user interacts properly with a programming *environment*. If the code is selected again (in case of evaluation of individual lines, placing the cursor in the relevant line is enough), and then re-evaluated, it will be executed again, and the post window will keep track of the new interaction. Let us discuss a possible source of confusion. On the one hand, the selected code is executed in sequence (the interpreter reads the selection "line by line", more correctly, as we will see, "expression by expression"), on the other hand the text document on which the user is working interactively is not exactly a page that has necessarily to be read from top to bottom. Rather, it is better to consider the text field as a blackboard on which the user is free to write wherever they wish, at the top, or at the bottom: the order of code execution is dictated by the order of evaluation. In this sense, what is left written in the text document is a memory of the user interaction through writing.

Back to the issue of the relationship between programming languages, a second aspect can be underlined in relation to binary vs alphabetical representation. One may argue that the latter is more abstract than the other, but this abstractive process does not necessarily ends here. In other words, you might

---

[1] Source code also includes SC code, which is accessible while using SC (see later).

think of another language in which the whole string "SuperCollider" is represented by a single character, such as @. This language would be at a higher level of abstraction, because its compilation/interpretation would output a representation (SuperCollider), which in turn should be interpreted/compiled. Without going deeper, the concept is that there may be different levels of abstraction with respect to the physical machine. Such a stratification is useful exactly because it allows the end user to forget about the physical machine itself (which is still there) by providing linguistic constructs, useful to model some specific conceptual domains (e.g. a certain musical structure: rhythmic, harmonic, melodic etc). The distinction between high and low level in the languages is usually related to the closeness/remoteness to/from the user. High-level languages are closer to "human" conceptual forms (and typically less efficient), low-level languages are nearer to technological aspects of the computer (and typically more efficient). SuperCollider is a high-level language, but relies for audio generation (a computationally intensive task) on a dedicated, low-level and very efficient component, the audio server (more on this later). In this way the overall SC application manages to hold together a high level of abstraction for the control (where the user directly works) and efficiency in computing audio. Finally, in the history and design of programming languages, many different conceptual paradigms have been proposed: in particular, the most accepted classification recognizes four different paradigms, imperative, functional, logical and object-oriented.

To summarize, SuperCollider is a programming language that has a wide general scope, but that has its strength and specificity in the description of notions related to electronic/digital audio and music. It is a high-level language that allows one to represent those same notions in a conceptually elegant way, by means of an advanced abstraction from their implementation. It is also a language that follows the paradigm of object-oriented programming. The latter is the focus of the next section.

2.2 Minima objectalia

Object-Oriented Programming (OOP) assumes that the user, in order to program the behavior of a computer, manipulates entities with properties and capabilities. The term –intentionally generic– indicating these entities is "objects", while their properties are typically thought of as "attributes" and their capabilities as "methods" that objects can adopt to perform operations.

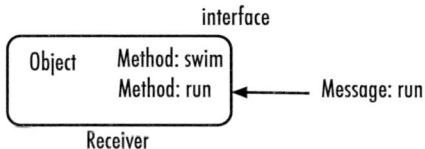

Figure 2.1   Structure and communication in OOP.

In the Object-Oriented paradigm the programmer deals with a world that presents itself as a set of objects that, under certain circumstances, may be manipulated. In particular, in order to manipulate an object – that is, to ask it to do something – the object must receive a "message", and conversely the programmer has to send it a message. An object, in order to respond to the message, needs to know an appropriate method. In short, the object can respond to a request (message) only if it has an appropriate competence (a method). In terms of communication, the object is the "receiver" of that message and it can answer if it implements a corresponding method. In summary:

- *object* and *method* concern the object *definition* from *inside*

- *message* and *receiver* concern the *communication* with the object from *outside*

The set of messages to which an object can respond is called its "interface": and it is an interface in the proper sense, because it makes the object available to the user for interaction, and the user can also be another object. The situation is depicted in Figure **2.1**. In most Object-Oriented languages, the typical syntax to send a message to an object uses the dot (.) and takes the form object.message. The relation between the receiver and the message must not be thought of as a description, like the verb would be in the third person ("the object does a certain thing"), rather as a couple of vocative/imperative forms: "object, do something!". For example, pencil.draw, or hippopotamus.swim. In short, to speak figuratively, the programmer is a kind of sorcerer's apprentice who tries to control a set of heterogeneous objects[2].

---

[2] Not surprisingly, the OOP paradigm is also considered as a specification of the imperative paradigm.

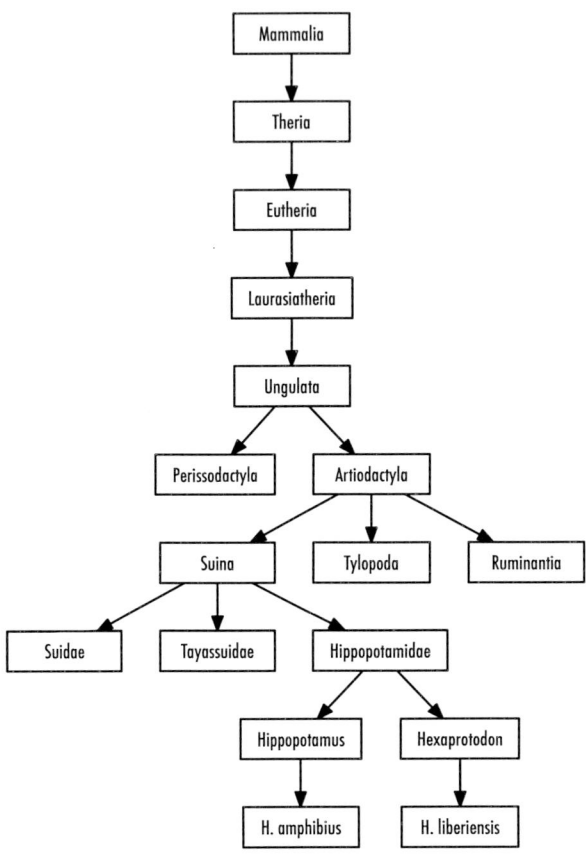

Figure 2.2   Taxonomy of the hippopotamus.

In order to be recognized by the language, the objects must belong to a finite set of types: an object is of type A, the other of type B, and so on. In OOP these types are called "classes". An object is therefore a particular "instance" of a certain class: the class can be thought of as the abstract type, but also as the "typeface" that produces instances. From a single typeface (the class) an indefinite number of characters can be printed (that is, the instantiated objects). As methods are defined in the class, all the instantiated objects of that class will be equipped with them. A class describes also the way in which an object is created ("construct") from the class itself.

Classes are organized hierarchically: each class can be derived from another class, and each class can have subclasses. This principle is called "inheritance". For example, a coinage is a further specification of a more generic "mold": the mold is the superclass of the coinage, and the coinage is a subclass of the mold. A seal (for sealing wax) is another mold, but of a completely different kind from the coinage: the seal is a subclass of the mold, from which it inherits some features that it shares with the coinage (the ability of printing), but that has some other specific features (the seal has a handle, while the coinage is supposed to beaten by a hammer-like device).

Inheritance can be thought of in a similar way to genetics, as features of the parent are present in the children (like in genetic heritage): but note that inheritance in OOP is of systematic nature, not evolutionary. The relation of inheritance relation finds a close model in natural taxonomies. For example, the graph in Figure 2.2 illustrates the taxonomic position of hippopotamus. The hippopotamus belongs to the Suborder of Suina (e.g. evidently, pigs), sharing with them some features, which differ both from Ruminantia (such as cows), although Ruminantia and Suina are both Artiodactyla (and distinguished both from Perissodactyla, such as horses). While classes are abstract (e.g. the species of hippopotamus), objects (good old hippos swimming in African rivers) are concrete.

The reason behind OOP is the application of a principle of encapsulation. Each class (and any object that derives from it) defines clearly and strictly its features in terms of data (attributes) and processes (methods). If it is a hippopotamus, then it can swim, but it cannot fly.

The reason behind inheritance is a principle of cognitive and computational economics. It would be useless to specify nursing with milk as a property of each single mammal species (hippo, man, hyena, etc.). Instead, it is useful to define such a property at an upper level in the taxonomic tree so that it is automatically passed on to all the lower nodes deriving from it.

2.3 Objects in SC

The SuperCollider language is an object-oriented language which adheres to OOP in a very "pure" form. It uses, as its historical model to which it is typological very close, the Smalltalk language[3]. In Smalltalk, as in SC, literally every possible entity is an object. This radicalness may be initially unsettling, although it is a strength, as it ensures that all (really all) the entities will be controlled by the user according to a single principle: all entities will have attributes and methods, and then it will be possible to send them all messages as they will expose a certain interface (i.e. a certain number of available methods) to the user.

An early example of particularly relevace is that of data structures: SC has many classes to represent data structures, entities that act as containers for other objects, each class being equipped with special skills and specialized for certain types of objects. For example, an "array", a fundamental data structure in computer science, is an ordered container of objects. Let us write Array. SC knows that Array is a class because the first letter is capitalized (see the next chapter): in SC everything starting with a capital letter indicates a class. If you evaluate the code, SC returns (for now this means: it prints to the screen) the class itself. Information about the class Array are available in the related help file, by selecting the text and choosing from the menu *Help* the Look Up Documentation for Cursor entry. After calling the help file, the reader will immediately see that many classifications of information are available, describing the relationship that Array has with other classes, its parent and its children. For example, it inherits from ArrayedCollection, which in turn inherits from SequenceableCollection, and so to on. The help file provides some guidance on the methods available for array-like objects. Let us consider the following code:

```
1  z = Array.new;
```

The example builds a new, empty array by sending the message new to the class Array[4]. So far, the assumption has been that a message is sent to a particular instance, not to class. But before you can send a message to an object, it is necessary that an object exists. All classes in SC respond to the message new,

---

[3] Actually, SC also includes aspects of other languages, primarily C and Python.
[4] Note that the code is colored according to the syntax: the classes as Array are (arbitrarily) assigned the blue color.

returning an instance. The method new is the "constructor" of the class: that is, the method that instantiates an object from the class (just like, "typeface, print me a character"). In addition to new, other constructor methods can be associated to a class, each returning an instance with some specific features: they are all *class* methods because they are sent to the class, not to the object. At this point the reader may speculate that, since in SC everything is an object, the class is also in some sense, an object, as it adheres to the OOP principle. In the case of Array, another constructor is the message newClear. It also includes a part between brackets:

```
1  z = Array.newClear(12) ;
```

In the example, the brackets contain a list of "arguments" (one, in this case), which further specify the message newClear, like in "object, do something(so)!"[5]. In particular newClear(12) provides an argument (12) indicating that the array will contain a maximum of 12 slots. It is possible to explicitly indicate the name of the argument, like in "object, do something (in the way: so)!". Every argument has a specific name, its *keyword*: in the case of newClear, the keyword for the argument is indexedSize, indicating the number of slots contained in the new array. The following code is the same as above, but notice how, in this case, it explicitly names the keyword for the argument:

```
1  z = Array.newClear(indexedSize:12) ;
```

Finally, z = indicates that the array will be assigned to the variable z. Note that the letter used is lowercase: if the reader wrote Z, SC will interpret Z as a (non-existent) class and raise an error. Now z is an empty array with a size of 12 slots and is an instance of the class Array. It is possible to ask z to communicate the class to which it belongs by invoking a method class:

---

[5] Or, back to our hippopotamus, hippo.swim(fast).

```
1  z.class
```

The method class returns the class of z: the evaluation of the code prints Array on the post window. Translating into plain English, the sentence would be: "z, declare your class". When using arrays, the user may often be dissatisfied with the methods listed on the help file file: many intuitively useful methods may seem to be missing. It is indeed one of the drawbacks: many times the desired method is present, but it is defined in the superclass and inherited by its subclasses. From the Array help file it is possible to navigate the structure of the help files going back to ArrayedCollection, SequenceableCollection, Collection: these are all superclasses (of increasingly abstract type) that define methods that are inherited by subclasses. Ascending along the tree, we reach its root Object. As stated in the help file:

> "Object is the root class of all other classes. All objects are indirect instances of class Object."

In other words, all the classes in SC inherit from Object, and therefore all objects are "indirectly" instances of Object. An example of inheritance is the method class that in the example above has been called on z: it is defined at the level of Object and inherited, by the relations specified by the class tree, by Array, so that an instance of that class (z) is able to respond to it.

Apart from navigation in the structure of the help files, SC provides the user with many methods to inspect the internal structure of the code: this feature is typically called "introspection". Methods such as dumpClassSubtree and dumpSubclassList print on the post window respectively a hierarchical representation of the subclasses of the class, and a list in alphabetical order. The two representations are equivalent. In the first, family relationships between classes through the tree structure are easier to spot, in the second it is possible to follow -for each of the subclasses of the class - the structure of the tree along the ascending branches up to Object. If we consider the class Collection –a very general class that has among its subclasses Array– and we send the messages dumpClassSubtree and dumpSubclassList, namely:

```
1  Collection.dumpClassSubtree ;
2  Collection.dumpSubclassList ;
```

this is what the interpreter prints on the post window in the two cases:

```
 1  Collection
 2  [
 3    Array2D
 4    Range
 5    Interval
 6    MultiLevelIdentityDictionary
 7    [
 8      LibraryBase
 9        [ Archive Library ]
10    ]
11    Set
12    [
13      Dictionary
14      [
15        IdentityDictionary
16        [
17          Environment
18            [ Event ]
19        ]
20      ]
21      IdentitySet
22    ]
23    Bag
24      [ IdentityBag ]
25    Pair
26    TwoWayIdentityDictionary
27      [ ObjectTable ]
28    SequenceableCollection
29    [
30      Order
31      LinkedList
32      List
33        [ SortedList ]
34      ArrayedCollection
35      [
36        RawArray
37        [
38          DoubleArray
39          FloatArray
40            [ Wavetable Signal ]

42  [...]

44  ]
45  Collection
```

```
 1  Archive : LibraryBase : MultiLevelIdentityDictionary : Collection : Object
 2  Array : ArrayedCollection : SequenceableCollection : Collection : Object
 3  Array2D : Collection : Object
 4  ArrayedCollection : SequenceableCollection : Collection : Object
 5  Bag : Collection : Object
 6  Collection : Object
 7  Dictionary : Set : Collection : Object
 8  DoubleArray : RawArray : ArrayedCollection : SequenceableCollection :
 9      Collection : Object
10  Environment : IdentityDictionary : Dictionary : Set : Collection : Object
11  Event : Environment : IdentityDictionary : Dictionary : Set : Collection :
12      Object
13  FloatArray : Raw

15  [...]

17  36 classes listed.
18  Collection
```

With Collection.dumpClassSubtree we see the position of Array in relation to its neighbors. It is on the same level of RawArray, both are subclasses of
ArrayedCollection. The latter class belongs to the family of SequenceableCollection. The method Collection.dumpSubclassList lists the classes in alphabetical order: it is easy to find Array, we can then follow the branches of the tree
(on the same line) up to Object.

Figure **2.3** is a visualization by means of a tree graph of part of the structure
of the classes Collection, obtained by automatically processing the output of
Collection.dumpSubclassList. The example is taken from the help file *Inter-
nal-Snooping*, which is dedicated to introspection in SC[6]. The last line printed
by SC is in either case the object Collection (39), which is what the methods
actually *return*. The reason behind this will be discussed shortly. By replacing
Collection with Object, the whole class structure of SC will be taken into account (and it's a big structure). Figure **2.4** shows a representation of the radial

---

[6] Also the help file for Class is particularly interesting in this respect.

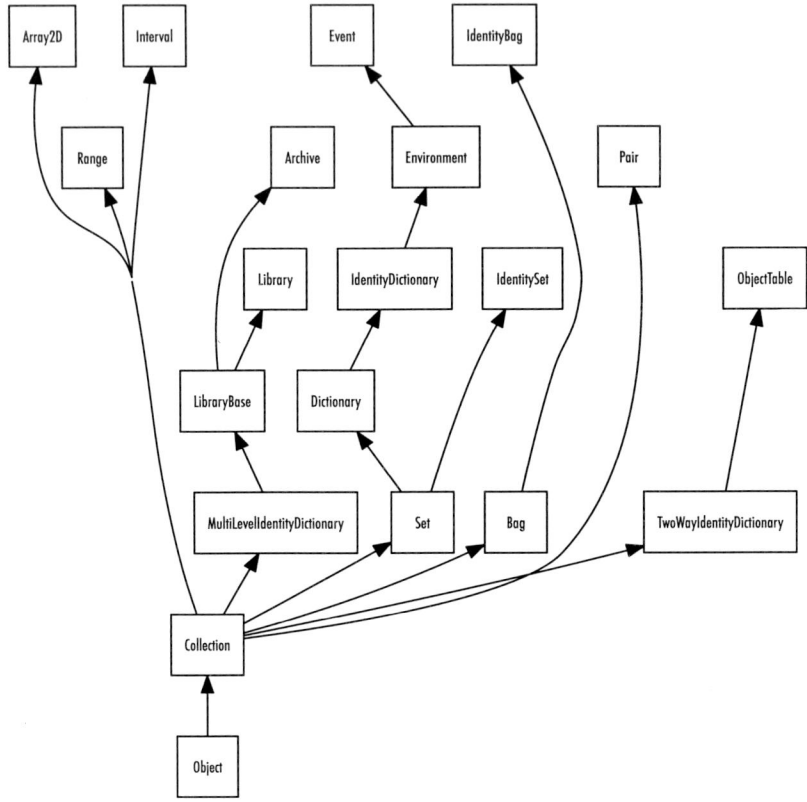

Figure 2.3   Tree structure for some of Collection subclasses, from Object.

structure of all classes of SC, obtained by processing the result of Object.dump-
SubclassList[7]. Note that a particulary thick point is represented by Ugen, the
direct superclass of all classes that generate signals in SC. Understandably, it is
a very large class.

_____

[7] Properly, the figure depicts the class structure installed on the machine of the
author, which includes some additional classes.

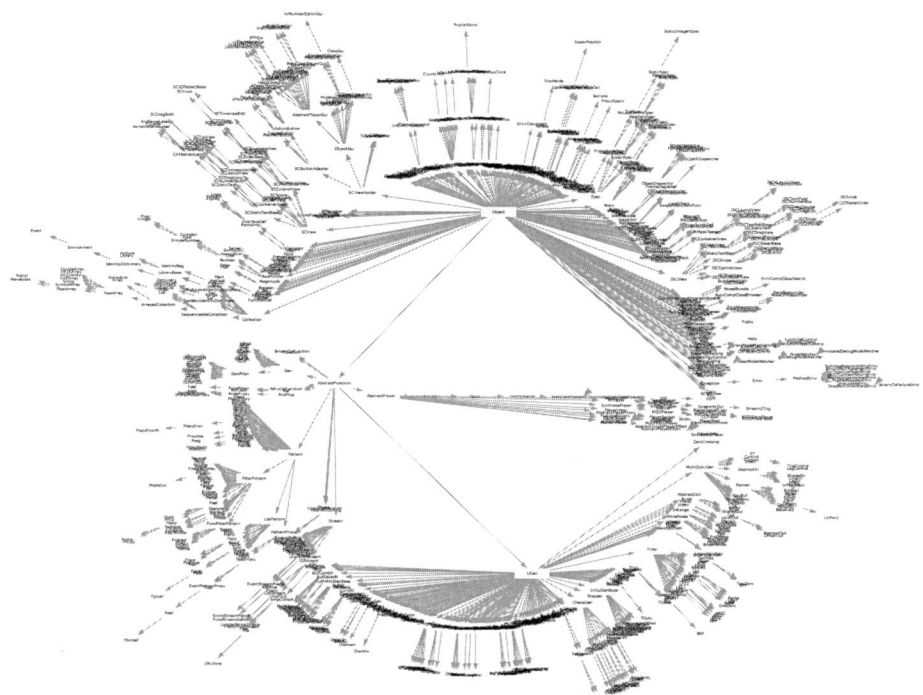

Figure 2.4 Radial graph for class structure in SC, from `Object`.

2.4 Methods and messages

The SC language is written for the most part precisely in SC (with the exception of a core of primitives written in C++ language for reasons of efficiency): thus, the same code that defines the language is transparent to the SC user. Of course, to be able to recognize the SC syntax, is very different from understanding exactly what it says: but in any case by looking at the sources it is possible to discover plenty of interesting information. It is the fact of being mostly written in SC that provides SC with a great power of introspection. After selecting `Array`, it is possible to access the definition of the class Array by selecting the `Look Up Implementations for Cursor` entry in the menu *Language*.

```
 1  Array[slot] : ArrayedCollection {

 3      *with { arg ... args;
 4          // return an array of the arguments given
 5          // cool! the interpreter does it for me..
 6          ^args
 7      }
 8      reverse {
 9          _ArrayReverse
10          ^this.primitiveFailed
11      }
12      scramble {
13          _ArrayScramble
14          ^this.primitiveFailed
15      }
16      mirror {
17          _ArrayMirror
18          ^this.primitiveFailed
19      }
20      mirror1 {
21          _ArrayMirror1
22          ^this.primitiveFailed
23      }
24  // etc
25      sputter { arg probability=0.25, maxlen = 100;
26          var i=0;
27          var list = Array.new;
28          var size = this.size;
29          probability = 1.0 - probability;
30          while { (i < size) and: { list.size < maxlen }}{
31              list = list.add(this[i]);
32              if (probability.coin) { i = i + 1; }
33          };
34          ^list
35      }

37  // etc
38  }
```

Without going into details, note that the first line (1) defines the class Array as a subclass of ArrayedCollection (Array[slot] : ArrayedCollection). By

(3) onwards, there is the list of the methods implemented for the class (width, reverse, scramble, each enclosed by a pair of curly brackets).

An easy way to retrieve a list of implemented methods is to use the power of introspection of SC. SC provides many ways to learn about information related to its internal state. The methods dumpInterface, dumpFullInterface, dump-MethodList display on post window information on the methods implemented as the interface for a class. In particular:

- dumpInterface: posts all the methods defined for the class;
- dumpFullInterface: as before, but also includes methods inherited from the superclasses;

The following example shows the result of evaluating Array.dumpInterface. The lists provided by the two other methods are much longer.

```
 1  Array.dumpInterface
 2      reverse (  )
 3      scramble (  )
 4      mirror (  )
 5      mirror1 (  )
 6      mirror2 (  )
 7      stutter ( n )
 8      rotate ( n )
 9      pyramid ( patternType )
10      pyramidg ( patternType )
11      sputter ( probability, maxlen )
12      lace ( length )
13      permute ( nthPermutation )
14      allTuples ( maxTuples )
15      wrapExtend ( length )
16      foldExtend ( length )
17      clipExtend ( length )
18      slide ( windowLength, stepSize )
19      containsSeqColl (  )
20      flop (  )
21      multiChannelExpand (  )
22      envirPairs (  )
23      shift ( n )
24      source (  )
25      asUGenInput (  )
26      isValidUGenInput (  )
27      numChannels (  )
28      poll ( interval, label )
29      envAt ( time )
30      atIdentityHash ( argKey )
31      atIdentityHashInPairs ( argKey )
32      asSpec (  )
33      fork ( join, clock, quant, stackSize )
34      madd ( mul, add )
35      asRawOSC (  )
36      printOn ( stream )
37      storeOn ( stream )
38      prUnarchive ( slotArray )
39      jscope ( name, bufsize, zoom )
40      scope ( name, bufsize, zoom )
41  Array
```

Through the same procedure that has been discussed to access the class definition (*Language* →Look Up Implementations for Cursor), it is possible to start from a method and trace the classes that implement it.

The following example, which could be a session with the interpreter, to be read from top to bottom by evaluating each line, allows us to move on in the discussion of methods.

```
1  z = [1,2,3,4] ;
2  z.reverse ;
3  z ;
4  z = z.reverse ;
5  z ;
6  z.mirror ;
7  z ;
8  z.reverse.mirror.mirror ;
```

The following is what appears in the post window as a result of the evaluation line by line:

```
1  [ 1, 2, 3, 4 ]
2  [ 4, 3, 2, 1 ]
3  [ 1, 2, 3, 4 ]
4  [ 4, 3, 2, 1 ]
5  [ 4, 3, 2, 1 ]
6  [ 4, 3, 2, 1, 2, 3, 4 ]
7  [ 4, 3, 2, 1 ]
8  [ 1, 2, 3, 4, 3, 2, 1, 2, 3, 4, 3, 2, 1 ]
```

However, a smoother way to create an array is simply to write the same array between square brackets (according to a widespread notation in programming languages)[8]. For example, z = [1,2,3,4] assigns the array [1,2,3,4] to the variable z (1). In the session with the interpreter the code z = [1,2,3,4] is evaluated (code, 1): SC returns the array [1, 2, 3, 4] (post window, 2) and

---

[8] This is a convenient abbreviation. SC language provides many of these abbreviations (called "syntactic sugar") that allow a gain of expressivity, but potentially introduce confusion in the neophyte.

assigns it to z. The last object returned from SC is printed on the post window as a result of the process of interpretation.

As we saw in the class definition above, one of the methods that the class Array provides is reverse: intuitively, the method takes the array and reverses the order of its elements. The moment that the message reverse is passed to z, the latter becomes its *receiver* (3). Then, z looks for the method reverse among those that are defined in its class, and it behaves accordingly. In the case in question, it is not relevant how the reverse operation is carried out by SC: in any case, by looking at the definition of the method (in class Array, page 30, lines 8-11), it is possible to see that the method includes a mysterious line, _ArrayReverse (line 9): the sign _ indicates that the reverse operation is carried out by a primitive of SC, written in the C++ language and not in SC. On the contrary, in the same class the method sputter (class Array, lines 25-35) is entirely written in SC. The methods return entities as a result of the operations that they carry out: these entities are objects in their own right. For example, z.reverse returns a *new* array, the reverse of z. On line 2, z.reverse asks z to carry out the operations defined by reverse: the result [4, 3, 2, 1] is not assigned to any variable (post window, 2). As we see, if z is called (code 3), the result is [1, 2, 3, 4] (post window, 3). In order to assign to z the result of the computation performed by reverse, the value calculated by the method must be reassigned to z, through z = z.reverse (code, 4). When called, z (code, 5) returns its value: this time it is the array z, now reversed (post window, 5). The method mirror instead generates a new array from that to which the message is passed, symmetrical to the center ("palindrome", so to say): z.mirror (code, 6) returns [4, 3, 2, 1, 2, 3, 4] (post window, 6), again without assigning it to z. The last line of code (8) highlights an important aspect of SC: the so-called "message chaining". The result of z.reverse is passed to mirror, the output of the latter is then passed to mirror again (post window, 8). These are the steps starting from z = [4, 3, 2, 1] (initial value plus three messages):

$$[4,3,2,1] \rightarrow [1,2,3,4] \rightarrow [1,2,3,4,3,2,1] \rightarrow [1,2,3,4,3,2,1,2,3,4,3,2,1]$$

Although message chaining allows one to write code concisely, it should be used with caution as it can make the code difficult to read. The next example shows two expressions that use chaining.

```
1  [1,2,3,7,5].reverse.reverse ;
2  [1,2,3,7,5].class.superclass.subclasses[2].newClear(3) ;
```

The first case presents no peculiarities compared to what has been discussed, but it is included for comparison to the second, the effects of which are very different. From [1,2,3,7,5] the first message class returns the class; then, the superclass of the class is accessed by superclass, again returning a class; from the superclass, through subclasses, an array containing all the subclasses is obtained (an array is just a container, so it can typically contains any object, here classes). The notation anArray[2] has not yet been introduced: the number 2 is an index (position) in the array. In essence, the method [n] allows to access the element $n + 1$ of the array. So, [Polynomial class, class RawArray, class Array][2] returns class Array. On the returned element (remember that is a class, Array) is then possible to invoke a constructor method that generates a new, empty array (nil) with three slots [nil, nil, nil].

A schematic representation of what happens in the two cases is in Figure **2.5**.

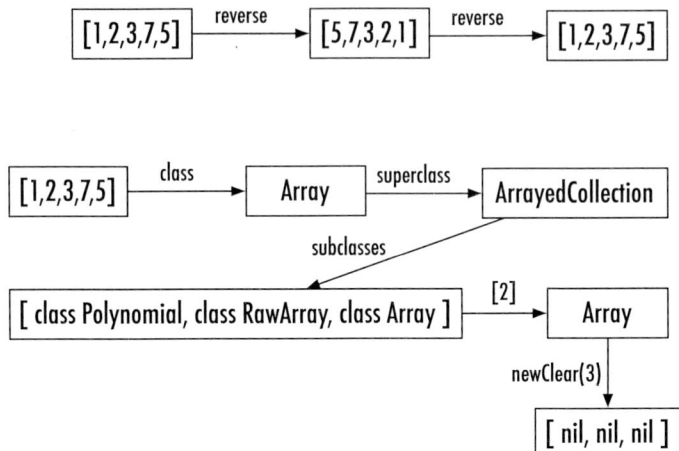

Figure 2.5 Two examples of message chaining.

2.5 The methods of type post and dump

So, all the methods always return an object. To avoid confusion we must remember the behavior of methods to obtain information through the post window. Examples already seen are dumpClassSubtree and dumpSubclassList, and dumpInterface, dumpFullInterface, dumpMethodList. The method that is most often used to obtain information, and inform us of what is happening, is postln, which prints a string representation of the object on which it is called and wraps to a new line. For example, let us consider the code Array.postln. Once evaluated, the expression prints on the post window:

```
1  Array
2  Array
```

When the method postln is called for Array, SC executes code that then prints the information for Array: in this case, Array being a class, it simply prints the class name, Array (1). At the end of each evaluation process, SC always

prints in the post window information about the last object that was returned
by an expression: in essence, SC calls postln on this object. In fact, the previous
expression returns again Array (2). In this case the utility of postln is virtually
nothing, but let us consider instead the following case:

```
1 z = [ 4, 3, 2, 1 ] ;
2 z.postln.reverse.postln.mirror.postln.mirror
3 z.postln.reverse.postln.mirror.postln.mirror.postln
```

The evaluation of the three expressions prints in the post window the three
following blocks:

```
1  [ 4, 3, 2, 1 ]

3  [ 4, 3, 2, 1 ]
4  [ 1, 2, 3, 4 ]
5  [ 1, 2, 3, 4, 3, 2, 1 ]
6  [ 1, 2, 3, 4, 3, 2, 1, 2, 3, 4, 3, 2, 1 ]

8  [ 4, 3, 2, 1 ]
9  [ 1, 2, 3, 4 ]
10 [ 1, 2, 3, 4, 3, 2, 1 ]
11 [ 1, 2, 3, 4, 3, 2, 1, 2, 3, 4, 3, 2, 1 ]
12 [ 1, 2, 3, 4, 3, 2, 1, 2, 3, 4, 3, 2, 1 ]
```

The array [4, 3, 2, 1] is assigned to z. The chained messages reverse.mir-
ror.mirror are called, but after each message a postln message is added (chained).
Basically, in this case postln lets the user see (in a visual form) the intermediate
result returned by each of the methods called. Note that it is useless to chain a
new postln after the last mirror (as in the code, line 3), since, as we saw, by default SC posts the result of the last computation (the results returned by mirror).
In fact, lines 11 and 12 produce identical results.

One might expect that since postln is used to print in the post window, the
method would return a string-type object, a set of characters. But *fortunately*, it
does not. In fact, postln:

- *prints* a string on the screen

• *returns* the object on which the method is called

These two things are completely different. The first is a behavior that is independent from computation, the second one concerns the computation flow, because the *returned* object is available for further processing. From this point of view (which is what matters for message chaining) postln is totally *transparent* with respect to the object on which it is called. This behavior is absolutely crucial in the process of code error checking ("debugging"), because it allows the user to chain postln messages to check the behavior of the methods called, but without interfering with the computation process. If the method postln would return a string, then in an expression, such as z.postln.reverse, the reverse message would be received by a string object and not by an array object, as happens in the next example, in which reverse is called on a string:

```
1 z = [ 4, 3, 2, 1 ] ;
2 z.postln.reverse.postln.mirror.postln.mirror
3 z.postln.reverse.postln.mirror.postln.mirror.postln
```

The result would then be:

```
1 ] 4 ,3 ,2 ,1 [
```

In other words, the reverse operation is applied to the characters that compose the string (see later for a detailed explanation). This type of behavior is typical for all SC printing and introspection methods. There are many methods that will print information in the post window: for example variants of postln are post, postc, postcln. For introspection, the reader is referred to the examples of Collection.dumpClassSubtree, Collection.dumpSubclassList, Array.dump-Interface. In all three cases, the last line prints in the post window the object returned by the method: note how the class is returned, as posted in the respective last lines (`Collection`, `Collection`, `Array`).

2.6 Numbers

The interpreter of SC can be used as a calculator. For example, consider this interactive session:

```
1  2.3*2 ;
2  4/3 ;
3  4**3 ;
4  4+2*3 ;
```

The interpreter responds on the post window with:

```
1  4.6
2  1.3333333333333
3  64
4  18
```

Two things stand out. The first one is very (very) important, and concerns the computation order. In the expression 4 + 2 * 3, unlike the standard mathematical convention, there is no hierarchy between operators: meaning, the multiplication is *not* evaluated before addition. In the expression, 4 + 2 is evaluated first, followed by * 3, which is calculated on the result of the previous operation $(4 + 2 = 6 \times 3 = 18)$. Computation order can be forced with the use of brackets, like this: 4 + (2 * 3). The second aspect that might impress the reader is that the syntax in use here contradicts the assumption that in SC everything is an object with an interface, so that each operation should follow the general pattern `object.method`. Here indeed SC makes an exception, at least for the basic four operations, which can be written more intuitively in the usual functional form. But this is just a convention of notation. That is, numbers (integers, floating point, etc.) are objects to all effects. If the message class is sent to an integer, for example 5 (1), the interpreter returns the class to which the instance belongs: Integer. Then, the method superclasses can be sent to Integer, that returns an array containing all the superclasses up to Object. Evaluated line by line, the code:

```
1 5.class ;
2 Integer.superclasses ;
```

returns:

```
1 Integer
2 [ class SimpleNumber, class Number, class Magnitude, class Object ]
```

Intuitively, without taking into account Object (the superclass of all classes), Magnitude is the class that more generally deals with magnitudes (including numbers). With Magnitude.allSubclasses we would get:

```
1 [ class Association, class Number, class Char, class Polar, class Complex,
2 class SimpleNumber, class Float, class Integer ]
```

Magnitude.dumpClassSubtree posts a tree representation of the subclasses of Magnitude: all classes that deal with magnitudes: Integer –the integer numbers– is near to Float –floating point numbers– since they are two subclasses of SimpleNumber. The latter class is part of a broader set of subclasses of Number –numbers in general, including polar and complex ones (Polar, Complex).

```
 1  Magnitude.dumpClassSubtree
 2  Magnitude
 3  [
 4     Association
 5     Number
 6     [
 7        Polar
 8        Complex
 9        SimpleNumber
10          [ Float Integer ]
11     ]
12     Char
13  ]
14  Magnitude
```

As they are objects (that is, properly, indirect instances of Object), it is possible to send to a number, 3 for example, the message postln, which prints the number and returns the number itself. So the code:

```
1  3.postln ;
2  3.postln * 4.postln ;
```

returns the following on post window:

```
1  3
2  3
3  3
4  4
5  12
```

Where (1) and (2) are the output of the code line (1), and (3), (4), (5) of the code line (2). The reader should remember what we discussed about the call by the interpreter of postln on the last evaluated expression and about the "transparency" of the latter method (that unobtrusively returns the same the object on which it is called).

For many mathematical operations, a double notation is available, functional and object-oriented[9].

```
1  sqrt(2) ;
2  2.sqrt ;

4  4**2 ;
5  4.pow(2) ;
6  4.squared ;

8  4**3 ;
9  4.pow(3) ;
10 4.cubed ;
```

For example sqrt(2) (1) prompts to compute the square root of 2: said in an OOP flavor, the method sqrt is invoked on 2 (2), returning the result of the square root applied to the object on which it is called. Similarly, the exponentiation can be written functionally with 4**2 (4), or as with 4.pow(2) (5): the pow method is called with the argument 2 on the object 4. Or, translated into natural language: "object 4, exponentiate yourself with exponent 2". Still, a third option (6) is to use a dedicated method for elevating to the square, squared. The same applies to the exponentiation to the cube (8-10).

2.7 Conclusions

The different paradigms of programming languages offer radically different ways of thinking about programming itself. The SuperCollider language adheres to the OOP paradigm, in many ways ensuring a strong and immediate conceptualization of the entities that are represented. Above all, the SC language is extremely consistent, and this aspect, once past the first difficulties,

---

[9] These are indeed other cases of "syntactic sugar".

makes the learning curve much smoother even for the novice. After this brief introduction to object-oriented programming, it is indeed appropriate to address in detail the syntax of SC, as well as write a real program.

3 Syntax: basic elements

As in any language, to speak SuperCollider you need to follow a set of rules. And, as in all *programming* languages, these rules are quite inflexible and binding. In SuperCollider, a sentence must be syntactically correct, otherwise it is incomprehensible to the interpreter, which means it will report it to you. This aspect may not be exactly friendly for the beginner not used to programming languages, forcing her/him to an accuracy of writing very far from an "analog" mood. However, there are at least two positive, interrelated, aspects in writing code. The first is the unavoidable analytical effort, which brings out a precise analysis of the problem to be solved, as the latter must be formalized linguistically. The second is a specific form of self-awareness: even if "bugs" in the language are possibile (although very rare), the programmer's mantra is: "If something does not work, it's your fault".

3.1 Brackets

In SuperCollider code examples, you will often find brackets, (), used as delimiters. Brackets are not expressly designed with this aim in the syntax of SC. However, it is a long-established writing convention (just look at the help files) that their presence indicates a piece of code that must be evaluated altogether (i.e., selection and evaluation of all lines). Remember that when you write or open a document, code evaluation is in charge to the user. In the IDE, double-clicking after an open bracket will thus allow the selection of the entire block of code until the next closing bracket: in this way, brackets facilitate a lot the interaction of the user with the interpreter. Brackets are also a way of arranging more complex code in blocks that must operate separately: for example, a block that must be evaluated in the initialization phase, another that determines a certain interaction with an ongoing process (audio or GUI), and so on.

3.2 Expressions

An expression in SC is a finished, complete sentence written in the language. Expressions in SC are closed (finished) by ; (a semicolon). Each code block enclosed by ; is therefore an expression of SC. The interpreter will collect the code until a ; is found, and then it will analyze the resulting expression. If the code to be evaluated by the interpreter consists of a single line (e.g. during an interactive working session), the ; can be omitted. In that case the interpreter will treat the entire block as a single expression (as it indeed is).

```
1  a = [1,2,3]
2  a = [1,2,3] ;
```

If evaluated line by line, the two previous expressions are equivalent. In general, it is always better to get used to including the ; even when you are working interactively with the interpreter, evaluating the code line by line. When evaluating multiple lines of code, the presence of ; is the only information available to the interpreter to know where one expression ends and another begins. In essence, when the interpreter is required to evaluate the code, it will start to scan the text character by character, and determine the closing of an expression in function of the ;. If all the code is evaluated all together (in one block) as in the following example (note also the use of brackets for delimitation):

```
1  (
2  a = [1,2,3]
3  a = [1,2,3]  ;
4  )
```

even if the intention of the user was to write two expressions, as far as the interpreter is concernered there is only one: being meaningless (not well-formed, to speak with Chomsky), the interpreter will report an error and will block the execution.

In the following example, the two expressions are the same because the line break is not relevant for SC. This allows you to use line breaks to improve the readability of code, or make it worse, of course, as happens in the following example.

```
1  (
2  a = [1,2,3]
3  )


6  (
7  a = [     1,
8            2,
9            3    ]
10 )
```

Again, note how in this case the absence of the semicolon in the second multiline version does not create a problem. In the absence of a ;, SC considers an expression as all that is selected; for the selected code is actually a single, well-formed expression, an so the interpreter does not report errors.

The order of expressions is their order of execution. In other words, the interpreter scans the code, and when it finds an expression terminator (; or the end of the selection) it executes the expression; then, it re-starts scanning thereafter until it has obtained a new expression to be executed.

3.3 Comments

A "comment" is a block of code that the interpreter does not take into account. When the interpreter reaches a comment indicator, it jumps to the end indicator of the comment, and it resumes normal execution process from then on. Comments are therefore meta-textual information that are very useful to make your code readable, up to the so-called "self-documenting code", in which the code also includes a guidance for its use. In SC there are two types of comments:

a. // indicates a comment that occupies one line or the ending part of it. When the intepreter finds //, it skips the following characters and jumps to the next line;

b. the couple /* ...*/ defines a multi-line comment: all the text included between the /* and */, even if it occupies more than one line, is ignored by the interpreter.

The following examples shows the usage of comments, in a very verbose way.

```
 1  /*
 2  %%%%% VERBOSE DOCUMENTATION %%%%%
 3
 4  There are 3 ways to get a power of 3
 5      - n**3
 6      - n.pow(3)
 7      - n.squared
 8
 9  %%%%% END OF VERBOSE DOCUMENTATION %
10  */

12  // first way
13  2**3 // it's 8
```

3.4 Strings

A "string" is a sequence of characters enclosed by double quotes. Strings can span multiple lines (remember that a line break is also a character in a string). A string is an ordered sequence of elements just like an array. In fact, the class String is a subclass of RawArray: strings are sequences of objects that can be accessed. Evaluating line by line the following code:

```
1  t = "string"
2  t[0]
3  t[1]
4  t.size
```

will output on the post window:

```
1  string
2  s
3  t
4  6
```

Here, t[0] indicates the first element of the array "string", that is: s, and so on.

It is the class String that defines all the post-related methods. When you send a post message an object, SC typically asks the object its string representation, and invokes the post method on the resulting string. For example, the method ++ for string concatenation works even if the objects to which the selected string are not of string-type: ++ asks internally to all the concatenated objects to provide a string representation: "you " ++ 2 is thus equivalent to "you " ++ (2.asString), and returns the string "you 2".

3.5 Variables

A variable is a placeholder. Whenever you store something, you assign it to a variable. In fact, to store the data in the memory is not enough: in order to access it, you must know its address, its "label" (as when you look for an object in a warehouse based on its location). If the data is stored but inaccessible (as in the case of an object somewhere in the warehouse), then you cannot use it and its presence is only a waste of space. The theory of variables is a very complex area in the science of computing. For example, one important aspect concerns the type of variables. In "typed" languages (for example, C), the user declares that he will use that label (i.e. the variable) to contain only and exclusively a certain type of object (for example, an integer), and the variable cannot be used for different objects (for example, a string). In such a case, before using a variable, its existence must be declared and its type must be specified. On the contrary, "untyped" languages do not require the user to specify the type of the variables, as the type is inferred in various ways (for example, by inspecting the object assigned to the variable). Some languages (e.g. Python) does not even require the declaration of the variable, which is simply used. In Python, the interpreter infers that that specific string represents a variable. Typing imposes constraints on the use of variables and a more verbose writing in code, but provides a clearer data organization. In untyped languages, writing code is faster, but on the other hand you can potentially run into complicated situations, such as when you change the type of a variable "on the run" without realizing it.

SuperCollider follows a sort of mixed approach. You must declare the variables you want to use, but not their type, which is inferred from their assigned object. In SC variable names must begin with a lowercase alphabetic character and can contain alphanumeric characters (uppercase characters, numbers). The variable declaration needs to be preceded by the reserved word var (which therefore cannot be used as a variable name). You can assign a value to a variable while you declare it.

```
1  (
2  var first, second;
3  var third = 3;
4  var fourth;
5  )
```

If you evaluate the example above, you may notice that the interpreter returns nil. In evaluating a block of expressions, the interpreter always returns
the last value: in this case, the one assigned to the variable fourth, which has
not yet been assigned a value, as indicated by nil. This word, nil, is another
"reserved word" that cannot be used as a variable name. Variable declaration
can also span multiple lines, provided that they are always consecutive in the
initial block of the code that uses them. In other words, variable declaration
must necessarily be placed at the beginning of the program in which they are
used. In SC there are two types of variables, *local* variables and variables related
to an *environment* ("environment variables"). As discussed, an interpreted language provides user interaction through an environment. In SC, environment
variables are variables that are constant (and thus usable) throughout the environment. Practically, they are permanently available throughout the working session with the interpreter. A particular case is given by the letters a - z
that are immediately reserved by the interpreter as environment variables. You
can use them (for example, in prototyping phases) without declaration. So far,
the examples involving variables have always referred to variables of this type,
such as a or z, as they allow interaction with the environment.

To understand the differences between the two types, consider the following example:

```
1  a = [1,2,3] ;
2  array = [1,2,3] ;
3  (
4  var array ;
5  array = [1,2,3]
6  )
```

```
 1  [ 1, 2, 3 ]
 2  ERROR: Variable 'array' not defined.
 3    in file 'selected text'
 4    line 1 char 17:

 6    array = [1,2,3] ;

 8  ---------------------------------
 9  nil
10  [ 1, 2, 3 ]
```

When evaluating line 1, a is a valid and legal name for an environment variable that does not have to be declared. The interpreter will then assign to a the
array [1,2,3] (post window, 1). When evaluating (2) the interpreter raises an
error (post window, 2-9), because it recognizes the assignment of a value to a
variable, but it notes that the local variable in question array has not been previously declared (· ERROR: Variable 'array' not defined.). The problem can
be solved by declaring the variable (code, 3-6; post window, 10). It is important
to note the use of brackets to indicate that the lines of code should be evaluated
together! The existence of the variable is valid only for the time at which the
code is evaluated. In other words, if you run again the expression array.postln,
you might assume that the variable array is declared and therefore legal. But
an error is raised by the interpreter. Local variables are therefore usable only
within those blocks of code that declare them. This is a good option for code
encapsulation, as there is no need to keep track of variables in use after the evaluation. On the other hand, during an interactive session it can be desirable to
maintain certain variables in existence to use them at a later time. Environment
variables are designed exactly for this purpose. The alphabetic characters a –
z are immediately assignable as environment variables by internal convention
of the intepreter. In addition, each variable whose first character is ~ (the tilde)
is an environment variable. Once declared in this way (*without* being preceded
by var):

```
 1  ~array = [1,2,3] ;
```

the variable ~array is set (i.e. fixed) throughout the session[1].

Users that are programming novices might wonder why not use only environmental variables. The answer is that they serve to work interactively with SC, to "communicate" –so to say– with SC, but not to write code in a structured form, as an example in designing systems for live music. In fact, in these contexts, environmental variables may be very dangerous for programming, simply because they are always accessible, and therefore poorly controllable.

To sum up:

- a variable name is an alphanumeric sequence of characters and always starts with a lowercase alphabetic character;
- variables are local or environmental;
- local variables must be declared at the beginning of the program and are valid only in its interpretation;
- environmental variables are permanent throughout the working session with the interpreter;
- environment variables are preceded by ~;
- individual alphabetic characters (not preceded by ~) are also predefined names for environmental variables.

3.6 Symbols

A symbol is a name that represents something in a unique way. It can be thought of as an absolute identification. It is a name that uniquely represent an object, a sort of proper name. A symbol is written in single quotes, or, if the character string does not include spaces within it, preceded by a \. The following example has to be evaluated line by line, and the result is shown on the window post.

---

[1] We will discuss further the issue of scoping, i.e. the scope of validity of the variables.

```
1 a = \symbol   ;
2 b = 'here a symbol' ;
3 a.class ;
4 [a,b].class.post.name.post.class ;
5 \Symbol .post.class.post.name.post.class ;
```

```
1 symbol
2 here a symbol
3 Symbol
4 ArrayArraySymbol
5 SymbolSymbolSymbolSymbol
```

Lines (1) and (2) assign to the variables a and b two symbols (1-2). Then, (3) asks to print as confirmation the class of a: Symbol. Finally, (4) is somewhat more esoteric. It use post, which prints but does not produce a line break, thus every print simply follows the previous one. The two symbols are placed in an array (the class returned by class): hence, the first Array. The codes asks for the name of the class (name). In fact, a class is an entity, and has a name attribute: naturally, the name for the class Array is Array. What is this name? It is a symbol that represents the class. The class of the *name* Array (that provides the name for the *class* Array) is Symbol. Let's rephrase it in a more figurative way: if the reader agrees to be a concrete example, whose class is Homo Sapiens. The label Homo Sapiens (the name of the class) is an object that belongs in turn to the class Taxonomic symbol. The last example (5) follows the same pattern (4) but the whole code is entangled with a little homonymy. The situation in both cases is shown in Figure **3.1**: its analytical understanding is left to the reader as an exercise.

Note the difference with strings. A string is a sequence of characters. For example, here a and b are two strings.

```
a = "symbol" ; b = "symbol" ;
```

The two strings are equivalent (roughly said: "they have the same content").

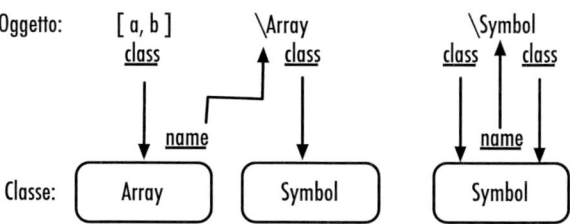

Figure 3.1 Relations between objects and classes in relation to Symbol.

```
1 a == b          // == asks: are they equivalent?
2 // post window replies: true
```

but they are not the same object (i.e. two copies of the same book are two different books).

```
1 a === b        // === asks on the contrary: are they the same object?
2 // post window replies: false
```

Instead, the relationship of identity is true in the case of the symbols:

```
1 a = \symbol ; b = 'symbol'     // same s
2 a == b; // post: true
3 a === b      // same object? post: true
```

Here, a and b are two containers for a unique object.

The previous examples introduce us to two other reserved words, true and false, dedicated to represent the values of truth. Reserved words (like var, nil and others to be seen later) cannot of course be used as variable names, even if their structure could be syntactically legal (they are alphanumeric and begin with lower case).

3.7 Errors

If you reconsider the previous example in which a variable was used without declaring it you can see how the SC interpreter reports errors:

```
1  ERROR: Variable 'array' not defined.
2     in file 'selected text'
3     line 1 char 17:

5     array = [1,2,3] ;

7  ------------------------------------
8  nil
```

As every computer interpreter/compiler, SC is very strict: it is a decidedly uncharitable interpreter. This requires special attention for beginners, who are likely to spend an interesting amount of time (beware) before being able to build a correct expression. In addition, error reporting is rather laconic in SC: in the above case it is very clear, in others it may be less clear. In particular it can be sometimes difficult to identify where the error is. Usually, the part of code indicated by SC while reporting the error is the point immediately next to, meaning just before, the error occurred. In this case what is missing is a variable declaration *before* = array [1,2,3].

3.8 Functions

Functions are one of the less intuitive aspects to understand in SC for those who do not come from a computer background. But let's consider the definition provided by the help file:

"A Function is an expression which defines operations to be performed when it is sent the 'value' message."

The definition is precise and exhaustive. A function is:

1. an expression
2. that defines operations
3. that are performed only when the function receives the message value. A function is therefore an object: it implements a method value which answers the value message (try Function.dumpInterface).

A function can be thought of as a (physical) object capable of doing certain things. For example, a blender. At the moment in which it is declared, SC is told to *build the object, but not make it work*. The object is then available: it is possible to make it work by sending it the message value. Function definitions are enclosed in curly brackets {}. Curly brackets are a sort of transparent shell that encloses the object, whose content is a set of expressions that will be evaluated when called. The concept of function is essential in structured ("organized") programming because it allows the programmer to apply a principle of *encapsulation*. Sequences of expressions that may be used multiple times can then be defined once, associated with a variable, and reused on demand.

```
1 f = { 5 } ;
2 f.value ;
```

The function f is an object that throws out on request the value 5. The definition stores the function object (1) whose behavior is triggered upon request using the message value (2). A more interesting use of functions, demonstrated by the following example, involves the use of "arguments": arguments can be thought of as input parameters for the object. Arguments are defined using the reserved word arg which is followed by the argument names separated by , and delimited by a ;. In the next example the function g is defined as { arg input;

input*2 }. Line (2): g takes one argument and returns the result of the operation on that argument. In particular g returns twice the value that input was
given. The function g is like a blender: you put the egg input and the blender
returns as its output (when set to operate) the result of blender.shake(egg)[2].

```
1  g = { arg input; input*2 } ;
2  g.value(2) ;
3  g.value(134) ;
```

```
1  A Function
2  4
3  268
```

Finally, the function h = { arg a, b; (a.pow(2)+b.pow(2)).sqrt } of the
last example is a calculation module that implements the Pythagorean theorem: it accepts in input the two catheti a and b and it returns the hypotenuse
c, according to the relation $c = \sqrt{a^2 + b^2}$. By the way, note that the square root
is implemented as a message sent to the integer resulting from computing the
bracketed operation.

```
1  h = { arg a, b; (a.pow(2)+b.pow(2)).sqrt } ;
2  c = h.value(4,3) ; // -> 5
```

In a second version (below) the definition does not substantially change but
allows to define additional aspects.

---

[2] The similarity with the syntax object.methodname is indeed not accidental.

```
1  (
2  h = { // calculate hypotenuse from catheti
3        arg cat1, cat2 ;
4        var hypo ;
5        hypo = (cat1.pow(2)+cat2.pow(2)).sqrt ;
6        "hypo: "++hypo } ;
7  )

9  h.value(4,3) ;

11 h.value(4,3).class ;
```

```
1  hypo: 5
2  String
```

Notice the following steps:

- comments work as usual within the functions (2);
- arguments must be specified first (3);
- argument names follow the criteria defined for variables (3);
- after the arguments, you can add a variable declaration (4). In the body of the function, especially if its complex, it may be useful to have some names for variables. In this case, hypo is a meaningful name that allows the last line to be more readable, where it is then used ("hypo:"++hypo). Of course, variables work as discussed before;
- a function returns a single value (be it a number, a string, an object, an array, etc.): the value of the *last expression* defined in the function body. The output of the last expression is then the output of the function. In particular, in this second version, the function h returns a string "hypo", to which (by means of ++) the content of the variable hypo is concatenated. Thus, in this case what is returned by the function is therefore a string. This aspects becomes evident if you evaluate the lines (9) and (10), the output of which is shown on the post window (below).

```
1  hypo: 5
2  String
```

This last point has important consequences. If you redefine h –in this first version– as proposed in the following example, it radically alters the way the function works.

```
1  h = { arg a, b; (a.pow(2)+b.pow(2)).sqrt ; a } ;
```

The addition of the expression a at the end of the definition means that the function h still computes the hypotenuse, but it returns as its output a (i.e. the first argument entered in (1)).

In short, a function has three parts, all three optional, but in the mandatory order:

1. a declaration of arguments (input)
2. a declaration of variables (internal operation)
3. a set of expressions (internal operation and output)

A function that has only the declaration of the arguments is an object that accepts some incoming entities, but litterally does nothing. In the following example, the function i accepts a in input, but the declaration of the argument is not followed by any expression: the function does nothing and returns nil. As an extreme case, a function that does not have any of the three components is still possible: in the example, the function l returns always just nil.

```
1  i = {arg a;} ;
2  l = {} ;
```

The situation may be represented as in figure **3.2**, where functions are represented as modules, which can be equipped with inputs and outputs. The text in the last row is the SC code relative to each diagram.

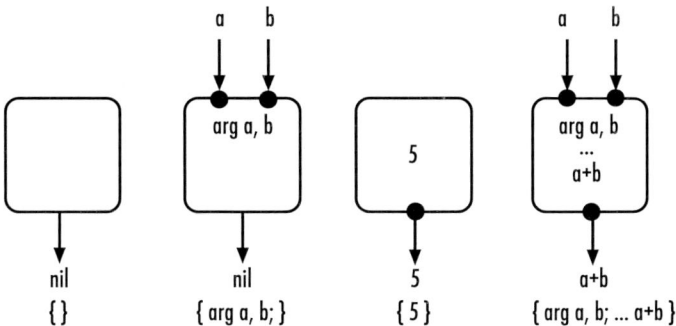

Figure 3.2 Functions.

By introducing functions it becomes possible to address the problem of the variable "scope"(their "visibility", "accessibility", so to say). Each variable "is valuable", it is recognized, and is associated to that value, and it is manipulatable, within a certain region of text: its "scope".

In the example below, func.value returns 8 because the variable val, having been declared outside the function func, is also visible inside the function.

```
1  (
2  var val = 4 ;
3  var func = { val*2 } ;
4  func.value ;
5  )
```

On the contrary, in the following example func always returns 8, as this now depends on the declaration of val inside func.

```
1  (
2  var val = 4 ;
3  var func = { arg val = 4 ; val*2 } ;
4  func.value ;
5  )
```

And so, the following example raises an error because val (declared in func) is not assigned a value, which effetively means that the operation is nil * 2, which is not legal.

```
1  (
2  var val = 4 ;
3  var func = { arg val ; val*2 } ;
4  func.value ;
5  )
```

In essence, the general rule is that the scope of variables goes from outside to inside. A variable is visible as long as the same name is not declared in a more internal block of code.

3.9 Classes, messages/methods and keywords

We have already seen how in SC, classes are indicated by a sequence of characters that begin with a capital letter. If you evaluate the following two lines of code (note the colored syntax):

```
1  superCollider ;
2  SuperCollider ;
```

you get for the two cases:

```
 1  ERROR: Variable 'superCollider' not defined.
 2     in file 'selected text'
 3     line 1 char 13:

 5     superCollider ;

 7  ------------------------------------
 8  nil
 9  ERROR: Class not defined.
10     in file 'selected text'
11     line 1 char 13:

13     SuperCollider ;

15  ------------------------------------
16  nil
```

As already mentioned, a message is sent to a class and to an object by means of the marker . : respectively, through the syntax `Class.method` and object.method. The methods can basically be thought of as functions defined for a certain class or object: when a method is invoked with a message it is like sending a message value to a function. In addition, methods may also have arguments as their input parameters. SC typically provides appropriate default values for the cases in which a method requires arguments, so that many times it may not be necessary to specify them. The use of keywords is useful because it allows to choose which is the desired argument that needs a certain value, leaving the other arguments with their default values. If keywords are not specified, the only criterion available for SC to assign a value to an argument is the order in which the value appears in the argument list. For example, the method `plot` for objects, which belongs to the class (and subclasses of) `ArrayedCollection`, includes the arguments

```
plot(name, bounds, discrete, numChannels, minval, maxval, parent, labels)
```

The method creates a window and draws in the form of a multipoint line the content of an object of type `ArrayCollection`. The argument name defines the title of the window. Thus, the window created by `[1,2,3,1,2,3].plot("test")` is labelled "test". The method also allows to define the number of channels numChannels. If the number is greater than 1, plot assumes that the first n samples are the samples no. 1 of the channels 1...n. For example, if there are two channels, then the first two samples are the samples number 1 of channel 1 and 2, and so on: plot draws a window for each channel[3]. In order to specify that numChannels must be equal to 2 without using the keyword, all the previous arguments have to be explicited. For example:

```
1  [1,4,3,2].plot("test", Rect(200 , 140, 705, 410), false, 2) ;
```

Much more easily, you can write:

```
1  [1,4,3,2].plot(numChannels:2)  ;
```

Finally, the use of keywords is in general slightly more expensive from a computational point of view, but it makes the code much more readable.

3.10 A graphic example

An example of code for creating a simple graphic element (GUI) allows us to introduce the fundamentals of SC syntax. The code below aims to create a graphic knob, that controls in parametric form the background color of a window.

SuperCollider comes with a rich palette of graphic elements that allow one to build sophisticated graphical user interfaces (GUI). So, although in SC the user interface in the implementation phase is indeed text-based, it does not mean that the same interface has to be used while performing, e.g. during a live setup. Obviously, the desired GUI must be programmed in SuperCollider.

---

[3] The utility of the method lies in the fact that multichannel audio signals, representable by an array, are stored in this "interlaced" form. If the signal is stereo, plot with numChannels: 2 draws the waveforms of both signals.

The case of GUI is a classic topic in object-oriented programming, because a GUI element is very well suited to be defined as an object (it is an entity clearly showing certain properties, e.g. colors, and behaviors, or actions related to its usage). GUI handling in SuperCollider is possible by including a widespread library, Qt, which is included within the IDE. Quite simply, it is immediately available for use. You will notice, however, that when using a GUI a second process is explicitly called, sclang, to which the GUI elements are related (also in relation to focus). For communication and representation, Qt GUI objects are represented in SC by means of classes: thus, Window is the SC class that represents a window object in Qt. As we are dealing with a language, it is like saying that the sign Window in SC is the conceptual and linguistic representation (a sign provided with an expression and a content) of a class of objects in the external world (a type of window in Qt). Let us look at what the code looks like:

```
1  (
2  /* coupling view and controller */

4  var window, knob, screen ; // declaring variables

6  // a container window
7  window = Window.new("A knob", Rect.new(300,300, 150, 100)) ;
8  window.background = Color.black ;

10 // a knob in the window, range: [0,1]
11 knob = Knob.new(window, Rect(50, 25, 50, 50)) ;
12 knob.value = 0.5 ;

14 // action associated to knob
15 knob.action_({ arg me;
16     var red, blue, green ;
17     red = me.value ;
18     green = red*0.5 ;
19     blue = 0.25+(red*0.75) ;
20     ["red, green, blue", red, green, blue].postln  ;
21     window.background = Color(red, green, blue);
22 });

24 // don't forget me
25 window.front ;
26 )
```

- **1**: the code block is enclosed in parentheses (1 and 26);
- **3**: a multi-line comment is used as a title (2) and there are other comments that provide some information about the different parts of the code (for example, 6). Commenting the code is of course optional, but is seen as good practice: it allows us to provide both general information on the structure and specific details about the implementation;
- **4**: apart from the comments, the code begins with the declaration of the three variables in use. They are declared at the beginning of the code block;
- **7-8**: the first thing to do is to create a container window, that is, a reference object for all other graphic elements that will be created later. This is a typical approach in the creation of GUI systems. The variable window is assigned an object Window, generated through the constructor method new. The method new is passed two arguments (i.e. new window, with these attributes): a string indicating the title of the window to be displayed ("A knob" and an object of type Rect. In other words, the window size, instead of being described as a set of parameters, is provided by a rectangle object, which is an instance of the class Rect, equipped with its attributes, that is, left, top, width, height. The position of the rectangle is given from top-left (arguments left and top), while width and height specify the dimensions. The rectangle then determines that the window will be 150×100 pixels, its upper left corner being placed in the pixel $(300, 300)$. Note that Rect.new returns an object (new is a constructor method) but *without* assigning it to a variable. In fact, on the one hand the Rect object does not need to be identifiabile outside of the GUI window, on the other hand, it actually remains accessible for future manipulations, since it is stored as a property of the window: bounds. This means, evaluating the expression window.bounds returns an object Rect. Following the same logic, that property can be modified, e.g. with this code: w.bounds = Rect (100, 100, 1000, 600), that assigns to window a new object Rect with different size and position as its bounds. Apart from the rectangle boundaries (so to speak), among the many attributes of the window there is the background color, accessible by calling the method background. The value of background can be set by passing an object Color. Even colors are objects in SC and the class Color provides some methods to easily obtain the most common colors by simply indicating their names, such as black with Color.black. Note that black is a constructor method that returns an instance of Color with the appropriate features for black;
- **10-12**: the construction of a graphic knob follows a similar procedure to that of the container window. The variable knob is assigned an object Knob (11). The contructor method works moslty like the one of Window: except that this

time it is also necessary to specify to which container window the knob must refer: the reference window here is indeed window, and the relative rectangle is defined by taking as a reference not the screen, but the window window. Therefore a rectangle 50×50, whose origin is in pixels $(30, 30)$ *of the Window* window. Note also that the geometry manager of Knob (as happened with Window) is obtained with the formula: Rect (50, 25, 50, 50). Now, here we have a class followed by a pair of brackets with arguments. However, where is the construct method (in this case .new)? This is an example of syntactic sugar. If after a class there are brackets with argument values, then the interpreter implies that the new method is implied. In short, new is the default method. That is, when the SC interpreter spots a class followed by a pair of brackets containing data, it assumes that you have invoked Class.new(arguments). The starting value of the knob is 0.5 (12). By default, the range of an object Knob varies between 0.0 and 1.0 (a *normalized* range): thus, by setting the attribute knob.value = 0.5, the knob is placed at half of its range. The ratio for having a normalized excursion lies in the fact that it is not possible to know in advance the use of the knob (will it control the frequency of an audio signal? $20 - 20,000$ Hz; or maybe a MIDI note? $0 - 127$). Note that the property is set through the use of the assignment operator =. Another syntax is available for setting attributes. In fact, the = is a synonym of the method *setter* represented by the symbol _, which explicitly assigns the value to the attribute. In other words, the two following syntaxes are equivalent.

```
1  knob.value = 0.5 ;
2  knob.value_(0.5) ;
```

The setter syntax is therefore of the type object.attribute_(value).
The next part of the code defines the interaction with the user.

- **15-22**: a knob is evidently a "controller", an object used to control something else. Therefore, it is possible to associate an action to the object knob: it is expected that by definition the action is carried out every time the user changes the value of knob. A function is appropriate for this type of situation, because it is an object that defines a behavior called from time to time and parameterized by arguments. The method knob.action asks to assign knob the action described by a function: the function definition is the code between the braces, 15-22. What happens behind the scenes is that, when the knob is moved, the function is sent a value message. The value message asks to evaluate the function for the value of the knob, the latter being the input of the function. In other words, the function answers the "What should be done when the knob knob moves?". In the function the input is described by the argument me (15): the argument has a completely arbitrary name (here chosen by the author), as it is used to represent in the reflexive form the object itself inside the function. This means the name me could be replaced with any other name (e.g. callOfCthulhu). Why the need for such an internal reference? As an example, it is useful to tell a graphic object to change *itself* as a result of its actions.

- **16-21**: in the example, however, the expected behavior requires changing the background color of window. Three variables are declared (red, green, blue) (16). They identify the three RGB components of the background color of window, that SC defines the range $[0, 1]$. The variable red is assigned the input value of me (17). The variables green and blue are assigned two values that are proportional (but in a different way) to me, so as to define a continuous change of the background for the three chromatic components in relationt to the value of me. This is a *mapping* operation: a certain domain of values ($[0, 1]$) is associated with three other domains, one for each component ($[[0, 1], [0, 0.5], [0.25, 1]]$). Then, the codes asks to print on the screen an array composed of a string and the three values (20), suitably formatted. Screen printing is not computationally unrelevant but it allows us to understand how values are calculated, and it is crucial in debugging. Finally, the attribute background of window is assigned an object Color, to which the three components are passed. The constructur of Color accepts that the three RGB components are specified in the range $[0.1]$. Again, Color (red, green, blue) is equivalent in every way to Color.new(red, green, blue).

- **25**: all GUI systems distinguish between creation and display. It is one thing to create GUI objects, and another to make them visible: this distinction allows to display/hide GUI elements on the screen without necessarily building and destroying new objects. The method front makes window and all the elements that depend on it visible: remember, without invoking it, all the code would work the same, but nothing would be displayed on the screen.

3.11 Control Structures

In SC the flow of computation follows the order of the expressions. Flow controls are those syntactic constructs that can change this order of computation. For example, a cycle for repeats the instructions nested inside it for a certain number of times, and then proceeds sequentially from there forward, while a conditional if evaluates a condition with respect to which the flow of information forks (wether the condition is true or false). Information flow controls are explained in the help file "Control structures", from which the following three examples are taken (with some small changes, respectively from if, while and for).

```
1  (
2  var a = 1, z;
3  z = if (a < 5, { 100 },{ 200 });
4  z.postln;
5  )


8  (
9  i = 0;
10 while ( { i < 5 }, { i = i + 1; [i, "boing"].postln });
11 )


14 for (3, 7, { arg i; i.postln });

16 forBy (0, 8, 2, { arg i; i.postln });
```

The first example shows the use of if. The syntax is:

```
    if ( condition to be evaluated,
  { function if the condition is true } ,
  { function if it is false })
```

In other words, the evaluation of the condition leads to a decision depending on the resulting true or false. Turning to the example, the variable a declared (note, it is local) is 1. The condition is a < 5. If the condition is true, the function { 100 } is computed, which returns 100; if it is false, the other function, { 200 }, is computed, returning 200. As the condition is true, the conditional statement returns a value of 100, which is assigned to z.

Also the control statement while has an evident semantics, borrowed from natural language

```
  while ({ condition is true }, { function to be computed } )
```

In the example, i is initially 0. As long as i is less than 5, the next function is called. The function increases i (otherwise the computation would loop in an infinite cycle) and prints an array that contains i and the string "boing".

Finally, the case of the cycle for, which iterates a function.

```
  for (start, end, { function } )
```

The function in the example is repeated five times $(3, ..., 7)$. The value is passed to the function as its argument so that it is available for computation: the function prints i at each call $(3, ..., 7)$. Note, the fact that the function's argument is named i is completely arbitrary (it is just a reference for internal usage). The two expressions of the following example shows that the position of the argument is the only relevant information that specifies the semantics (i.e. the counter), not the name (arbitrary):

```
1 for (3, 7, { arg i; i.postln });
2 for (3, 7, { arg index; index.postln });
```

The instruction forBy requires a third parameter that indicates the step:

```
  forBy (start, end, step, { function } )
```

The example is a variation of the previous one, that prints the range $[0, 8]$ every 2. There are indeed other control structures. Here it is worth introducing do, which iterates the elements of a collection. It can be written in this way:

```
do ( collection, function )
```

but more typically it is written as a method defined on the collection, i.e.:

```
collection.do({ function })
```

the example is taken from the help file "Control-structures", with some minor changes.

```
1  [ 101, 33, "abc", Array ].do({ arg item, i; [i, item].postln; });
2
3  5.do({ arg item; ("item"+item.asString).postln });
4
5  "you".do({ arg item; item.postln });
```

By evaluating the first line, the post window prints out:

```
1  [ 0, 101]
2  [ 1, 33 ]
3  [ 2, abc ]
4  [ 3, class Array ]
5  [ 1, 2, abc, class Array ]
```

To avoid any doubts, the last line simply returns the original array. The function is passed the element on which to perform the iteration (item) and a counter (i). Again, the names items and i are totally arbitrary (and they can be replaced with any other desired names). It is their place that specifies the semantics, i.e.

```
1  [ 101, 33, "abc", Array ].do({ arg moby, dick; [dick, moby].postln; });
2  [ 0, 101 ]
3  [ 1, 33 ]
4  [ 2, abc ]
5  [ 3, class Array ]
6  [ 101, 33, abc, class Array ]
```

The function prints an array that contains the counter i (left column of the
first four rows) and the element items (right column). The method do is also
defined in integers, its semantics this time being n times.evaluate function,
so to say. The way it works is shown in the second example. Evaluating the
code results in:

```
1  item 0
2  item 1
3  item 2
4  item 3
5  item 4
6  5
```

The last line is the integer on which the method is called. The function
prints a string consisting of the concatenation of "item" with a representation
as a string of the integer item (returned by the method AsString called on the
item) $(0, ..., 4)$. Since most of the cycles for iterate from 0 and with a 1 step, they
are often written in SC with do called on a number (e.g. 3.do). The syntax of do
(object.method) is fully OOP. Finally, the last example simply shows that each
string is a collection whose elements are individual alphanumeric characters.

3.12 Yet another GUI example

The following example shows the code to create a GUI which chooses a color,
Simple Color Selector. The control of GUI elements is a particularly interesting

way to demonstrate some aspects of the syntax, and the visualization of the process can be a useful aid.

The designed GUI consists of three sliders. Each slider controls one of the three parameters that define the color (the fourth, not considered here, is transparency, alpha). The color is displayed as the background of the window. In addition, the value obtained for each component is displayed in numerical form on the side of the relative cursor. The GUI allows one to choose between two typical color spaces. As seen, the color is defined in a computer through the RGB model. However, this model is not very intuitive compared to the perception of color: among others, a different mode is in use, describing color in terms of hue (going circularly from red to yellow to green to blue to purple, to red again), saturation (expressing the amount of color, from absence to full color), brightness (or "value", which indicates the brightness and goes from white to black). The model is therefore called "HSB" or "HSV". HSV encoding can be converted with a formula into RGB (the latter being the only format actually implemented on a computer). So, in SC Color provides a constructor hsv that allows us to define a color precisely through hue, saturation, value (always in the normalized range $[0, 1]$). In the GUI, a button is included to choose between RGB and HSV encoding. Clearly, there are various ways to arrange such a GUI, and still more ways to implement it in SC.

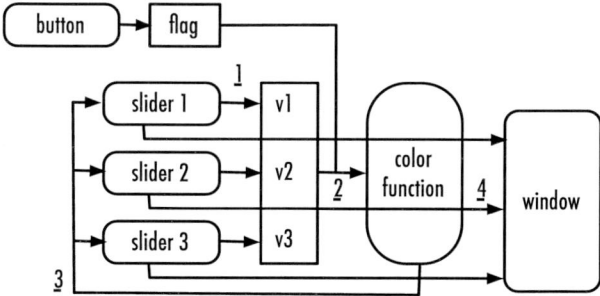

Figure 3.3   Structure of the Simple Color Selector.

A diagram of the project is shown in Figure **3.3**. In the example two elements can be recognized: a flag, which is a variable that can have discrete values (here two: in the code, rgb or hsv, and an array of three values (in Figure, v1-v3) to specify the color. A button allows us to choose which of the two values of the flag is active (and then the selected color space). The three sliders perform each four actions (numbered and underlined in the example). Each slider selects a relative value in the array (1), then calls the color function and passes it the array and the flag (2); the function returns a color (3), and change the background (4). The function has an important role because it is the computational core of the code, and it is kept separate from the user interface to ensure encapsulation. Therefore, the function will not refer to external variables, but will be passed them as arguments. Similarly, the function will not change the background, rather it will simply return the resulting color.

A possible implementation follows:

```
1  (
2  /*
3  Simple Color Selector
4  RGB-HSV
5  */

7  var window = Window("Color Selector", Rect(100, 100, 300, 270) ).front ;
8  var guiArr, step = 50 ;
9  var flag = \rgb , colorFunc ;
10 var colorArr = [0,0,0] ;

12 colorFunc = { arg flag, cls;
13     var color, v1, v2, v3 ;
14     # v1, v2, v3 = cls ;
15     if(flag == \rgb ){
16         color = Color(v1, v2, v3)
17     }{
18         color = Color.hsv(v1.min(0.999), v2, v3)
19     } ;
20     color ;
21 } ;

23 Button(window, Rect(10, 200, 100, 50))
24 .states_([["RGB", Color.white, Color.red], ["HSV", Color.white, Color.black]])
25 .action_({ arg me; if (me.value == 0) {flag = \rgb } {flag = \hsv } });

27 guiArr = Array.fill(3, { arg i ;
28     [
29         Slider(window, Rect(10, (step+10*i+10), 100, step)),
30         StaticText(window, Rect(120, (step+10*i+10), 120, step))
31     ]
32 }) ;

34 guiArr.do{|item, index|
35     item[0].action_{|me|
36         item[1].string_(me.value) ;
37         colorArr[index] = me.value ;
38         window.background_(colorFunc.value(flag, colorArr));
39 }} ;
40 )
```

- **7-10**: variable declaration. Note that window is immediately associated with the creation of the window that appears. Other variables are initialized with "meaningful" values;
- **12-21**: the block is dedicated to the definition of the function colorFunc. The function takes as input arguments flag and cls. The first is the flag, the second an array of the three color values. Line 14 introduces another useful abbreviation: # v1, v2, v3 = cls is equivalent to v1 = cls[0]; v1 = cls[1]; v3 = cls[2]. The conditional block (15) operates in relation to the verification of the flag. If flag is rgb (literally, if flag == rgb is true), then color is assigned a color according to the RGB color model (default). Otherwise (the cases are only two, RGB or HSV, so there are no other possibilities), the same values define a color built in reference to HSV. In the latter case, note that by definition (see the help file for Color.hsv) the argument hue can be set at a maximum of 0,999. If the first slider is moved to 1, then there will be a problem. The solution is the method min defined on numbers, which returns the lesser between the number on which it is called and a threshold value. So, for all values less than 0,999, it will return the value itself, while it will returns 0.999 if the number is greater. Note that the last expression of the function is simply an invocation of color, so that it will be returned in the output. Finally, note the "syntactic sugar" in the definition of the conditional statement, that increases readability by eliminating some brackets;
- **23-25**: a button is created. Note that, as it will be no longer changed, it is not associated with a variable for future reference (the object simply has to work). Of course, a button is built following the usual conventions that we discussed with other GUIs. In addition, it has a property, states, that defines each button state by a set of arrays (in SC, buttons are not limited to two states, they can have n). The parameters are easily inferable from the running GUI. The method action associates an action to the button (it is the usual semantics). But in this case, the value of a button is the index of the state: that is, the button has a value 0 in the first state, 1 in the second, and so on. Therefore, it becomes possible to define a condition statement depending on the state of the button. In this case, values are only two (as they have to be associated with the two possible values of the flag, through a symbol), and an if statement is enough, determining the value of the flag according to the state of the button;
- **27-32**: it is time to build sliders and the relative labels that show their values. The approach presented here is "oversized" for the case, but its aim is to be useful in a more general sense. Rather than building the six elements one

by one, a procedural approach is applied. The constructor method `fill` of `Array` returns an array of n places: for each place, an elements is provided by calculating the function passed as an argument to `fill`. The function has an argument that represents the counter (in the case, `i`). In our example, the array has a size of three elments, and the function returns for each place an array made of a slider and a label. Each element of the array `guiArr` is in turn therefore an array of two elements. The construction of the slider (`Slider`) and the label (`StaticText`) is very similar to what we saw for Knob. Note that the position of the graphic elements depends on a parameter `step` common to both the elements and controlled by the counter. The idea is "do it three times, a slider and a label, every time with the same size but progressively shifting down a bit". When `i = 0`, then the cursor is on the abscissa at the point 10, when `i = 1` is at 70, and so on. A similar approach is extremely useful in cases where the elements are not three but many more, or when the number of elements cannot be known in advance but could depend on some variable. The work involved in programming a GUI is compensated by its flexibility;

- **34-39**: the block defines the action of each slider. The action is assigned by iterating on the array that contains the elements. Each element is accessible through `item` and its position through `index` (the reader should remember that the argument names are arbitrary, its their position that defines their semantics). Now, every element of `guiArr` is an array with cursor and label. Therefore, `item[0]` will return the slider, and `item[1]` its relative label. So the action associated with each slider (35, the function associated with each movement of the slider) will consist of updating the value of the label associated (through its attribute `string`) (36); updating the array `colorArr` for the index `index` with the value of the slider (37); changing the background of `window` with the result of the function call `colorFunc` which is passed `flag` and `colorArr` (38). Note that this passage includes the steps 2-4 of Figure **3.3** (in which, by the way, labels are not included).

3.13 Conclusions

What we have seen in this chapter is a good start for diving smoothly into the SC language. There is much more, of course. But with some general references

it is possible to explore the language itself by inspecting the interactive help
files and the examples that they provide, by exploiting the internal snooping
that SC provide, and also by looking directly at the SC code source. It is worth
closing the chapter by mentioning some abbreviations (some "syntactic sugar")
that are useful and often used, but which may cause some confusion in the SC
novice:

```
1  // omitting new in Object
2  a = Something.new(argument) ;
3  a = Something(argument) ;

5  // omitting value in Function
6  function.value(aValue) ;
7  function.(aValue) ;

9  // multiple assignment to Array
10 # a,b,c = array ;
11 a = array[0]; b = array[1]; c = array[2] ;

13 // less parentheses
14 if (a, {do}, {do something different}) ;
15 if (a) {do} {do something different} ;

17 Something.methodWithFunction_({}) ;
18 Something.methodWithFunction_{} ;

20 // argument shortened by |
21 { arg argument, anAnotherArgument ; } ;
22 { |argument, anAnotherArgument| } ;
```

For a complete discussion the reader is referred to the help file "Syntax
Shortcuts".

4 Synthesis, I: Fundamentals of Signal Processing

SuperCollider is undoubtedly specialized in real-time sound synthesis by means of its audio server, scsynth. The aim of this chapter is not, however, to introduce the audio server and its functions. The following discussion aims instead to provide a rapid introduction to digital signal processing and sound synthesis using the SuperCollider *language*. The expert reader eager to learn about synthesis via the server can safely skip to the next chapter. The discussion on the fundamentals of synthesis still allows us to look more closely at some of the linguistic aspects previously introduced. Of course, the chapter is very concise, and the reader is referred to the many available resources on acoustics and digital audio.

4.1 A few hundred words on acoustics

A sound is a continuous variation of pressure detectable by the human ear. As a vibration, it depends on the movement of bodies of the physical world (a guitar played with a plectrum, a tent shaken by the wind, a table beaten with knuckles). A sound is thus a series of compressions and rarefactions of air molecules around the listener: what is propagated is precisely this oscillation (as in a system of steel balls that collide), not the molecules themselves, that instead oscillate around an equilibrium position. A signal is a representation of a time pattern, sound being a good example of this. In particular, an audio signal, that represents a sequence of compressions/rarefactions of the atmospheric pressure, takes the form of an oscillation between positive values and negative values. If this oscillation is regular in time, the signal is periodic, otherwise it is aperiodic: most sounds are placed between these two extremes, i.e. they are more or less periodic/aperiodic. The most basic periodic signal is a sine wave, something that acoustically corresponds more or less to the sound of a tuning fork. Figure **4.1** allows to summarize what has been said. Let us suppose one strikes the tuning fork (as a singer would do to get a reference note). Following the supplied energy, the tuning fork will produce sound, the latter resulting from the oscillation of the metal prongs. In other words, the oscillation of the metal prongs around the equilibrium position will be transmitted to the air molecules around it, which will fluctuate accordingly, "analogically". Therefore, the "form" of the oscillation of the prongs is substantially the same "form" as the oscillation of the air molecules, that the listener perceives as a sound. Suppose then to connect a pen to one of the prongs of the tuning fork and to have a tape that runs at a uniform speed (in fact, the model is precisely that of a seismograph): such a device will record the excursion of the prongs and the resulting trace, which is, in the case of the tuning fork, a sinusoid.

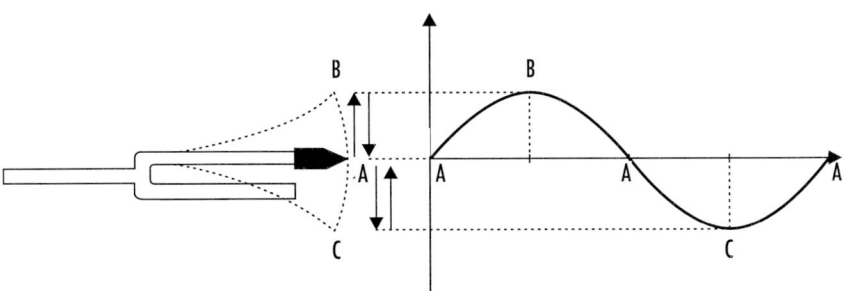

Figure 4.1   Vibration of a tuning fork and sinusoid.

By observing the sine wave, it is possible to see the basic dimensions of a signal:

- after *A* has been reached for the second time, ideally the trace is repeated. The signal is then *periodic* because it repeats its beahvior after a certain *period* of time, which is referred to by *T*. The *frequency*, *f* of a signal is the number of repetitions of the cycle in the time unit. Intuitively, $f = \frac{1}{T}$, and vice versa.

Audible frequencies are (approximately) in the range $[16, 20000]$ Hz (*Herz*). As the measure represents the number of cycles per second, it is also referred as *cps*;

- the *amplitude* of a signal is the amount of oscillation, in the example the maximum travel of the reed. Note that the audio signal is *bipolar*, i.e. has a positive and a negative maximum (the sign is arbitrary but represents the opposite directions of the oscillation). An audio signal is (usually) symmetrical with respect to the 0, that in turn represents the point of inertia (in which the reed is at rest). The amplitude measurement can take place in various ways. Very often (that is, in SC and in many other softwares) two units are in use. On the one hand, a normalized range $[-1, 1]$, abstract from the physical value, on the other hand a representation in *decibels, dB*, a unit of measure for sound intensity/pressure somewhat closer to the perception ;

- let us suppose to have two identical tuning forks that are hit one after the another after a certain amount of time. Intuitively, they will draw the same trace, but at each instant one will be in a certain position, the other in a different one. That is, they will have a difference of *phase, ϕ*. Since in periodic signals the cycle is repeated, the phase is measured in *degrees* (such as on a circumference) or *radians*, that is, in fractions of π, where 2π is the circumference;

- the oscillation motion of the tuning fork follows a certain trace, that is, the signal exhibits a certain *waveform*, the sine wave. Other acoustic sources (an oboe, for example) would result in different plots.

The above discussion concerns the so-called time-domain representation of a signal. The signal is in fact represented as a temporal phenomenon. An alternative representation of the signal is possible, in the frequency domain, in which the signal is represented with respect to its frequency content. In particular, the Fourier theorem states that any periodic signal can be decomposed into a sum of sinusoids of different amplitude: as if a (theoretically infinite) number of sinusoids of different volumes all sounded together. The sine wave is mathematically the simplest form of periodic curve (which is why sinusoids are often referred to with the terms "simple signal" and "pure tone"). The result of this decomposition is a *spectrum*, which can be seen as the way in which energy is distributed among the various sinusoidal components in which the input signal is decomposed. A spectrum does not include time information: it is like an instantaneous snapshot of the internal composition of the sound. Therefore, if we analyse a signal in the frequency domain we get its spectrum. Starting from the Fourier theorem, we can observe that the spectrum of a complex signal (not of a sinusoidal tone) is made up of many components of different frequency. In a periodic signal these components (called "harmonics") are integer multiples of the fundamental frequency (which is the greatest common divisor). Signals of this type are for example the sawtooth wave, the square wave, the triangular wave, and in general the stationary phases of all signals with a recognizable pitch. In an aperiodic signal, the components can be distributed in the spectrum in an arbitrary manner. When we are talking (very loosely) of "noise", we mostly refer to aperiodic signals.

4.2 Analog vs. digital

A digital signal is a numerical representation of an analog signal and is doubly discrete: it represents discrete amplitude variations (*quantization*) into discrete instants of time (*sampling frequency*, or better *rate*). The digitalization of an analog signal is shown in Figure **4.2**. A continuous signal (for example, a voltage variation of the electrical output from a generator which produces an analog sine wave at 440 Hz) (a), is sampled (b): in the sampling process, an analog-to-digital converter (ADC) is polling at regular time intervals (defined by a "clock") the voltage value. In essence, it imposes a "vertical" reference grid on the analog signal. The result (c) is a pulse signal (that is, it consists of pulses) between which there is no other information. A similar mechanism occurs in relation to the amplitude of the signal (d). Here the grid defines a set of discrete (finite) values to which the amplitude values detected in the analog signal have to be approximated (e). The result is a pulse signal whose amplitudes are reffered to a discrete scale (f): a digital signal.

The digitalization can be thought of as a grid that is superimposed on an analog signal, even if a digital signal is typically represented in software as a continuous curve for sake of convenience. Sometimes it is shown as a broken line: at the end, there is only one mathematically continuous function passing through all points, and therefore, once converted back to analog (through a digital-to-analog converter, DAC), the curve should have exactly the same shape as the input. Obviously the process produces approximations. Quantization defines the maximum dynamic range of the signal (in CD-quality, 16 bit, 96 dB) while sampling defines the maximum frequency that can be represented in the digital signal. As shown by the Nyquist theorem, the sampling rate states that

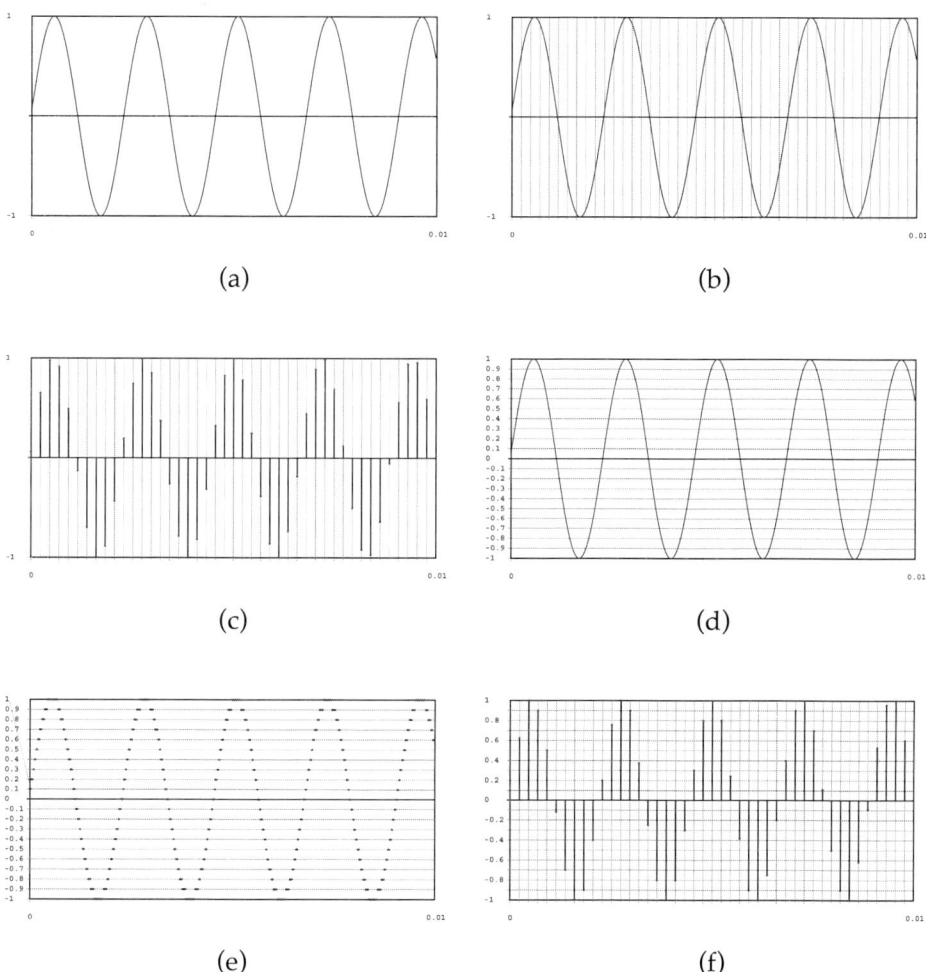

Figure 4.2 Digital signal: analog-to-digital.

the frequencies that can be represented in a digital signal are all those comprised in the half of the sampling frequency: if we have a CD-quality signal with a sample rate of 44.100, then we can represent frequencies up to 22.050 Hz (a good approximation of the audible frequencies).

In the case of digital audio, the digitalized signal is thus available in the form of numerical information. This can be processed by a computer which can perform the calculation, and process the numbers representing the signal. The main steps of digital recording are as follows (Figure **4.3**):

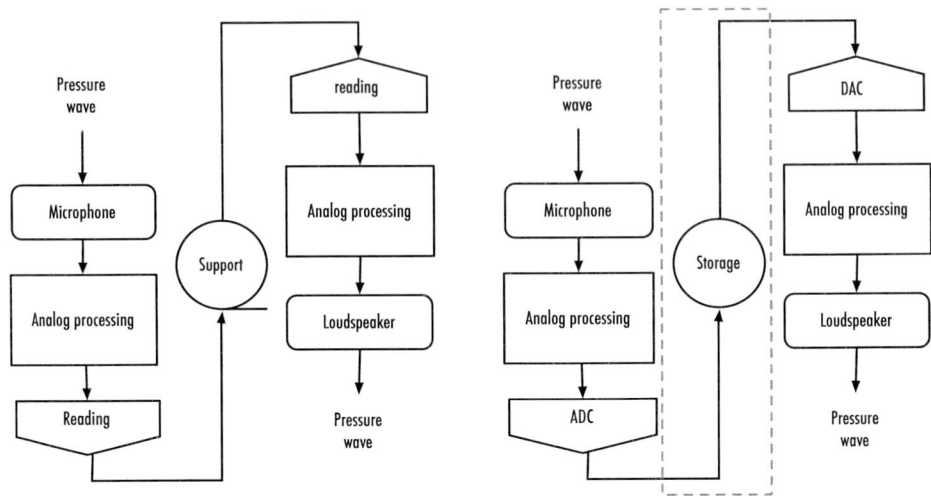

Figure 4.3   Audio chain: analog and digital.

1. **analog-to-digital conversion**: the analog signal is filtered and converted from the analog domain (as a continuous variation of the electrical voltage, e.g. produced by a microphone) in the digital format via the ADC;

2. **processing**: the digitalized signal, now a sequence of numbers, can be processed in various ways;

3. **digital-to-analog conversion**: to be heard, the sequence of numbers that composes the signal is converted back into an analog signal through the DAC: once filtered, the DAC generates again a continuous variation of the electrical voltage that can, for example, drive a loudspeaker.

In the preceding discussion, the process of digitalization has been described with respect to a pre-existing analog signal. On the other hand, the analog is always "surrounding" the digital. To be heard, a digital signal must necessarily be converted into a continuous electric signal, which is then sent to a speaker, making it vibrate. However, a digital signal does not have to start from an analog one. The starting assumption of computer music is that the computer can be used to directly synthesize the sound. The heart of digital synthesis is to exclude the step 1, directly generating the sequence of numbers that have to be converted into an analog signal. Such an assumption in no way precludes the possibility of working with samples from "external" sources, rather it underlines that the fundamental aspect of computer music lies in the methods and calculation procedures that govern the synthesis. It is always possible for the "digital" composer to work on analog components (e.g. recording and processing analog signals), but the most distinctive aspect of the composer's practice is to exploit the numerical (and therefore "computable") nature of digital signals (Figure **4.4**).

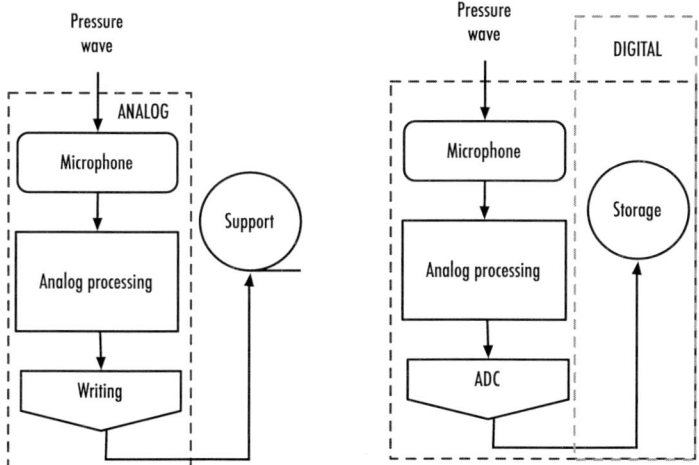

Figure 4.4   Analog vs. digital composition.

A discrete signal can be described through a mathematical function:

$$y = f[x]$$

the function indicates that for each discrete moment of time x the signal has the amplitude value y. A digital signal is a sequence of elements $x_0, x_1, x_2, \ldots$ which correspond to values of amplitude $y_0, y_1, y_2, \ldots$.

The data structure that represents a signal is typically an array, a sequence of consecutive and homogeneous memory cells, that is, the cells contain the same data type (the numeric type chosen for the signal in question). Thus, for example an array such as:

```
[0, 0.5, 0, -0.25, 0, 0.1, -0.77, 0.35]
```

The array describes a signal composed of 8 samples (Figure **4.5**), where the index (the number that label progressively each of the eight values) is x, to be

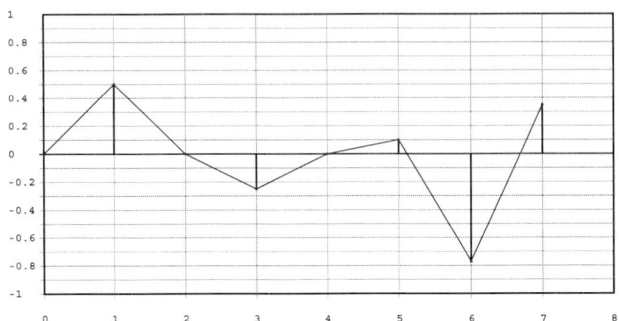

Figure 4.5   The array [0, 0.5, 0, -0.25, 0, 0.1, -0.77, 0.35].

considered as a moment of time, while the associated numeric data is y, defined as the amplitude of the signal at the time x:

$$x = 0 \rightarrow y = 0$$
$$x = 1 \rightarrow y = 0.5$$
$$\dots$$
$$x = 7 \rightarrow y = 0.35$$

SuperCollider allows us to easily view such data structures through the method plot. For example, the class Array provides the plot method that generates a window and draws the resulting curve by joining the values in the array.

```
1  [0, 0.5, 0, -0.25, 0, 0.1, -0.77, 0.35].plot ;
2  [0, 0.5, 0, -0.25, 0, 0.1, -0.77, 0.35].plot
3    ("an array", minval:-1, maxval:1, discrete:true) ;
```

The first line makes use of the default values, the second line gives an example of some possible options. The method plot is implemented not only in the class Array but in many others, and is extremely useful for understanding the behavior of the signals taken into account.

4.3 Synthesis algorithms

An algorithm for the synthesis of sound is a formalized procedure which has as its purpose the generation of the numerical representation of an audio signal.

The SuperCollider language (sclang) allows us to experiment with algorithms for signal synthesis in deferred time without taking into account –for the moment– the audio server (scsynth). This is not the usual way to use SuperCollider. Moreover, sclang is designed as a high-level language ("far from the machine and close to the programmer"), and it is inherently not optimized for massive numerical operations ("number crunching", as it is defined in computer science jargon). However, for educational reasons and in order to increase our knowledge both on signal synthesis and on the language itself, it can be useful to discuss some simple synthesis algorithms.

Since the signal of CD audio (currently still the most popular audio standard) is sampled at 44.100 Hz, in order to create a mono signal of 1 second with CD quality, we need to build an array of 44.100 slots: the computation process to generate a signal requires us to design and implement an algorithm to "fill" each of these (indexed) places with a value. So, to generate a pure sinusoidal signal, the easiest method is to calculate the amplitude y for each x sample of the signal, according to the sine function and to associate the value to y to the index x in the array A, that represents S.

A periodic function is defined as follows:

$$y = f(2\pi \times x)$$

A sinusoidal signal is described by the formula:

$$y = a \times sin(2\pi \times k \times x)$$

The effect of the parameters a and k is represented in Figure **4.6**, where a (discrete) signal composed 100 samples is drawn (in continuous form).

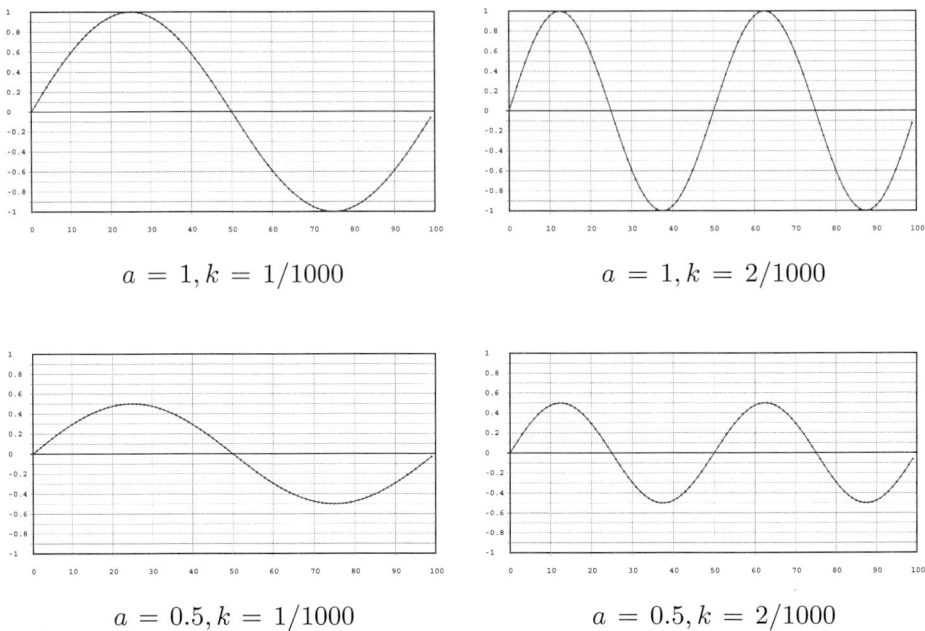

$a = 1, k = 1/1000$

$a = 1, k = 2/1000$

$a = 0.5, k = 1/1000$

$a = 0.5, k = 2/1000$

Figure 4.6 Sinewave and variation of parameters a and k.

The synthesis algorithm for a sinusoidal signal, written in pseudo-code (that is, in a non-existent language that allows us to illustrate programming in an abstract form), is the following:

```
For each x in A:
    y = a*sin(k*x)
    A[x] = y
```

Line 1 of the cycle calculates the value y as a function of two parameters a and k that control the amplitude and frequency of the sinusoid, while line 2 assigns index x of A the value y. SC allows to easily implement a similar algorithm. For example, the first signal of Figure **4.6** was obtained with the following code:

```
1  var sig, amp = 1, freq = 1, val ;
2  sig = Array.newClear(100) ;
3  sig.size.do({ arg x ;
4      val = amp*sin(2pi*freq*(x/sig.size)) ;
5      sig[x]= val ;
6  }) ;
7  sig.plot(minval:-1, maxval:1, discrete:true) ;
```

In the code, the first line defines the variables that contain respectively the signal, the amplitude, the frequency, and the incremental value of the samples. Then, an array of 100 elements is created. Lines 3-6 are occupied by a loop: sig.size returns the size of the array (100). For sig.size (100) times the function is evaluated in the cycle do: because x represents the increase along the array (representing time: $0, 1, 2...98, 99$), the function ($f[x]$) calculates its value in relation to x. The value of the frequency of the desired signal indicates the number of cycles (2π) per second. The case freq = 1 indicates that a cycle of the sine wave (ranging from 0 to 1) will be distributed over 100 points. Hence, the meaning of x/sig.size. If the desired frequency is 440 Hz ($\rightarrow$ cycles per second, freq = 440) then there should be $440 \times 2\pi$ cycles per second (2*pi*freq). This value should be distributed on all the places of the array (x/sig.size). The computed value of val is in the range $[-1, 1]$ by trigonometric definition and therefore can be directly scaled by amp. Line 5 assigns the value val to the place x of sig. The signal is then drawn in the amplitude range $[-1, 1]$, in a discrete fashion (7).

A class particularly useful for calculating signals is Signal, which is a subclass of ArrayedCollection (the superclass of many array-like objects) that is specialized in signal generation. The class Signal is designed to contain large arrays with homogeneous type of data (as typical for audio signals). The following example is an obvious rewriting of the code discussed above using Signal and filling a signal of 44.100 samples, equivalent to one second of audio at CD standard sampling rate.

```
1  var sig, amp = 1, freq = 440, val ;
2  sig = Signal.newClear(44100) ;
3  sig.size.do({ arg x ;
4      val = amp*sin(2pi*freq*(x/sig.size)) ;
5      sig[x]= val ;
6  }) ;
```

The resulting signal can be stored on the hard disk in an audio format in order to be played back, by means of the class Soundfile (still on the client side). In this way, Signal can be used to generate audio materials to be used later.

```
1  (
2  var sig, amp = 1, freq = 440, val ;
3  var soundFile ;

5  sig = Signal.newClear(44100) ;
6  sig.size.do({ arg x ;
7      val = amp*sin(2pi*freq*(x/sig.size)) ;
8      sig[x]= val ;
9  }) ;

11 soundFile = SoundFile.new ;
12 soundFile.headerFormat_("AIFF").sampleFormat_("int16").numChannels_(1) ;
13 soundFile.openWrite("/Users/andrea/Desktop/signal.aiff") ;
14 soundFile.writeData(sig) ;
15 soundFile.close ;
16 )
```

In the example, Soundfile creates an audio file (11) with properties defin-able by the user (12): file type ("AIFF"), quantization (16 bits, "int16"), number of channels (mono, 1)[1]. It is important to specify the quantization because SC internally (and by default) works with a 32 bit quantization in a float format: a useful format for the internal precision but rather inconvenient for final release. After the file object is created, the user must also specify a path where the file

---

[1] Note the chaining of messages: each of the setter methods returns the object itself.

is to be stored (13). At this point the data in the array sig can be written in the file (14) - data is generated exactly as in the previous example. Once the writing operations are concluded, the file must be closed (15), otherwise it will not be readable. The file is ready to be played back.

To avoid writing the generated data onto a file, we can use the method play that offers a chance to "directly" hear the contents of the Signal (how this happens will be discussed in details later) .

```
1  var sig, amp = 1, freq = 441, val ;
2  sig = Signal.newClear(44100) ;
3  sig.size.do({ arg x ;
4      val = amp*sin(2pi*freq*(x/sig.size)) ;
5      sig[x]= val ;
6  }) ;
7  sig.play(true) ;
```

What is the frequency at which the signal is played? Until now the rate has in fact been specified only in terms of relative frequency between the components. With play we switch to real-time[2], other variables have to be taken into account, that will be discussed later. SC by default generates a signal with a sampling rate (often shortened to sr) of $44,100$ samples per second.

The content of the array, after being put in a temporary memory location (a "buffer") is read at a frequency of $44,100$ samples per second (a signal of $44,100$ samples is "consumed" in a second). In other words, SC takes a value from the buffer every $1/44,100$ seconds. With the method play(true) (the default) the execution is looping: once arrived at the end of the buffer, SC restarts from the beginning. So if $size$ is the number of elements in the array, the signal period ("how much it lasts" in seconds) is $size/sr$, and the frequency is its inverse: $1/size/sr = sr/size$. If $size = 1,000$, then $f = 44,100/1000 = 44.1$ Hz. Conversely, if we want to get a signal whose fundamental is f, the size of the array that contains a single cycle must be $size = sr/f$. The calculation is only approximate because $size$ has to be an integer. If the array size is $44,100$ (as in many examples in the chapter, but not in all), then the signal is read once per second. Because the array contains a number of cycles $freq$, freq is actually the signal

---

[2] It is therefore necessary to boot the audio server, From the menu Language > Boot Server. Everything will be clarified later on.

frequency. In the following example(s) play will not be automatically used: the possibility of using the method is left to the reader, together with the task of adapting the examples.

Returning now to synthesis-related issues, a periodic signal is a sum of sine waves. A "rudimentary" implementation might be the following:

```
1  var sig, sig1, sig2, sig3 ;
2  var amp = 1, freq = 1, val ;
3  sig = Signal.newClear(44100) ;
4  sig1 = Signal.newClear(44100) ;
5  sig2 = Signal.newClear(44100) ;
6  sig3 = Signal.newClear(44100) ;

8  sig1.size.do({ arg x ;
9      val = amp*sin(2pi*freq*(x/sig.size)) ;
10     sig1[x]= val ;
11 }) ;
12 sig2.size.do({ arg x ;
13     val = amp*sin(2pi*freq*2*(x/sig.size)) ;
14     sig2[x]= val ;
15 }) ;
16 sig3.size.do({ arg x ;
17     val = amp*sin(2pi*freq*3*(x/sig.size)) ;
18     sig3[x]= val ;
19 }) ;
20 sig = (sig1+sig2+sig3)/3 ;
21 sig.plot ;
```

In the example we want to calculate the fundamental and first two harmonics. Four Signal objects are generated, with the same size (3-6). Then, four signals are calculated, by repeating a code that is structurally always the same, as it varies only for the presence of a multiplier freq (8-19). Finally, sig is used to contain the sum of the arrays (implemented for array as the sum of the elements in the same relative positions). The aim is precisely to obtain a "sum" of sinusoids. The values in the array sig are then prudently divided by 3. In fact, if the signal must be contained within the range $[-1, +1]$, then the worst possible scenario is that of the three peaks (be they positive or negative) are in phase (together), the sum of which would be just 3. Dividing by 3, then the maximum amplitude possible will be -1 or 1. The signal can be written onto an audio file as in the previous example. Note that, for graphic clarity, in order to draw onto the screen it is convenient to reduce the number of points and set $freq = 1$.

In programming, repetition of substantially identical code blocks is always suspect. In fact, iteration is not an ideal solution (even if it seems a simple solution). Moreover, the repetition typically results in errors. Furthermore, the program is not modular because it is designed not for the general case of a sum of n sinusoids, but for the specific case of three sinusoids. Finally, the code becomes unnecessarily verbose and difficult to read.

The following algorithm provides a more elegant way to calculate a periodic signal with a number of harmonics n, defined by the variable harm.

```
1   var sig, amp = 1, freq = 440, val ;
2   var sample, harm = 3 ;
3   sig = Signal.newClear(44100) ;
4   sig.size.do({ arg x ;
5       sample = x/sig.size ;
6       val = 0 ;
7       harm.do{ arg i ;
8           harm = i+1 ;
9           val = val + (amp*sin(2pi*freq*(i+1)*sample) );
10      } ;
11      sig[x]= val/harm ;
12  }) ;
```

The cycle (4) calculates the value of each sample and use sample to keep in memory the location. For each sample it resets val to 0. At this point, for each sample it performs a number n of calculations, so that it simply calculates the value of the sine function for the fundamental frequency and the first $n-1$ harmonics, as they were n different signals. At each calculation, it adds the value obtained to the one calculated for the same sample by the other functions (the "sum" of sine waves). Finally, val is stored in the element of sig on which we are working, after dividing it by the number of harmonics (following the cautionary approach discussed above). Compared to the previous example, the code uniquely defines the required function and assigns parameters as desired (try to vary harm increasing harmonics), it is easier to fix, it is general and more compact. It is worth introducing another point, although it will not have practical consequences. The iterative example implements an approach that is not meant to work in real time. First the three signals are calculated, then they are

summed (mixed, we may say). In real time, as will be seen, the signal is generated continuously, therefore it is not possible to follow such an approach as the
duration of the signal is not determined. On the contrary, the second example
potentially can be implemented in real time: in fact, the algorithm calculates
the final value of a sample one at a time. In this case, this value is written into
the array sig by increasing progressively the position, but it could instead be
sent to the sound card and converted to sound.

Starting from the last example, it is easy to generate other periodic signals,
in the same way as the ones already mentioned. By definition, a sawtooth wave
is a periodic signal that has theoretically infinite frequency harmonics $f \times n$,
where f is the fundamental frequency and $n = 2, 3, 4, ...$, and amplitude respectively equal to $1/2, 3, 4...$ (i.e. each one inversely proportional to the relative harmonic number). The following example introduces a small change in
the algorithm presented above.

```
1   var sig, amp = 1, freq = 440, val ;
2   var ampl ; // ampl is the same for each component
3   var sample, harm = 10 ;
4   sig = Signal.newClear(44100) ;
5   sig.size.do({ arg x ;
6       sample = x/sig.size ;
7       val = 0 ;
8       harm.do{ arg i ;
9           harm = i+1 ;
10          ampl = amp/harm ;
11          val = val + (ampl*sin(2pi*freq*(i+1)*sample) );
12      } ;
13      sig[x]= val/harm ;
14  }) ;
```

The only difference is the introduction of the variable ampl: for each sample, for each harmonic component (10) the relative amplitude is calculated by
dividing the reference amplitude amp by the harmonic number. By increasing
harm, it can be seen that the sawtooth wave is gradually approximated more
accurately.

A square wave can be generated in the same way as the sawtooth wave, but
adding only the odd harmonics ($n = 1, 3, 5...$). In other words, a square wave is
a sawtooth wave in which the even harmonics have zero amplitude. The code
is shown in the following example.

```
1  var sig, amp = 1, freq = 440, val ;
2  var ampl ; // ampl is the same for each component
3  var sample, harm = 20 ;
4  sig = Signal.newClear(44100) ;
5  sig.size.do({ arg x ;
6      sample = x/sig.size ;
7      val = 0 ;
8      harm.do{ arg i ;
9          harm = i+1 ;
10         if(harm.odd){ // is it odd?
11             ampl = amp/harm ;
12             val = val + (ampl*sin(2pi*freq*(i+1)*sample) );
13         }
14     } ;
15     sig[x]= val/harm ;
16 }) ;
```

This time the value of ampl is subject to a conditional evaluation (10), where harm.odd returns true or false depending on whether the variable (the harmonic number) is even odd or even. If it is odd, the value is calculated as in the case of the sawtooth wave, if it is even, the computation of val is skipped for the component.

Variable numbers of harmonics (3,5,40) for the three cases discussed (sinusoids with constant amplitude, sawtooth wave and square wave) are shown in Figure **4.7**. In particular, in the sawtooth and square waves the contribution of harmonics is made evident by the number of "humps" in the signal.

4.4 Methods of Signal

The class Signal provides many opportunities for processing that have not been discussed before, as a "direct" implementation was more useful for learning purposes. Signal is in fact a class that is used to generate signals that typically are not to be written to files, but that are to be used by specific synthesis methods (*wavetable* synthesis, to be discussed later). For example, the method sineFill

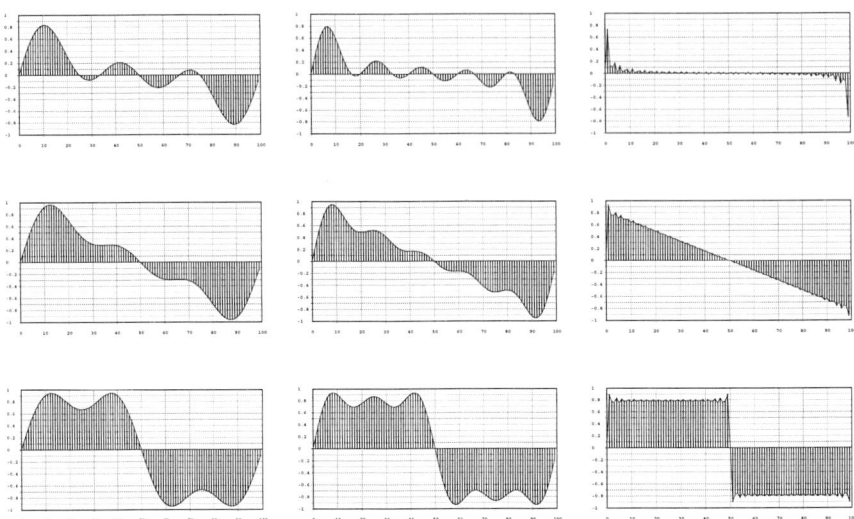

Figure 4.7 From left, the sum of 3, 5, 40 sinusoids: from above, constant amplitude, sawtooth wave, square wave.

is explicitly dedicated to the generation of signals by summing sinusoids. Its arguments are

1. the size;
2. an array specifying a series of amplitudes;
3. an array that specifies a series of phases.

In the following minimal example, a sinusoid of 100 points is generated by specifying a single element (i.e., the fundamental) in the array of amplitudes. Note also the method `scale` defined on `Signal` but not on other subclasses of `ArrayedCollection`, which multiplies the elements by a specified factor.

```
1  var sig, amp = 0.75, freq = 440 ;
2  sig = Signal.sineFill(100, [1]).scale(amp) ;
3  sig.plot(minval:-1, maxval:1) ;
```

As we said, amplitudes and phases are related to the harmonics of the sinusoidal signal. For example, an array of amplitudes [0.4, 0.5, 0, 0.1] indicates that the first 4 harmonics will be calculated, where F_2 will have amplitude

0.4, F_3 0.5, and so on. Note that to remove a harmonic component, its amplitude must be set to 0 (the case of f_4). The help file includes the code:

```
1  Signal.sineFill(1000,  1.0/[1,2,3,4,5,6])  ;
```

The code creates an array of 1000 points which it fills with a sine wave, plus its first five harmonics. The syntax 1.0/[1,2,3,4,5,6] is interesting. If further evaluated, the post window returns:

```
1  1.0/[1,2,3,4,5,6]
2  [ 1,  0.5,  0.33333333333333,  0.25,  0.2,  0.16666666666667 ]
```

That is, a number divided by an array returns an array where each element is equal to the number divided by the relative element of the starting array. In other words it is like writing [1.0/1, 1.0/2, 1.0/3, 1.0/4, 1.0/5, 1.0/6]. The array therefore contains a series of 6 amplitudes inversely proportional to the number of harmonics. Intuitively, the resulting signal approximates a saw-tooth wave[3].

In the following example, the approximation is much better. The method series, defined for Array, creates an array and has as arguments size, start and step: the array is filled by a series of size successive integers that, starting from start, increase by step. Therefore, the array contains the values $1, 2, 3, ...1,000$. The signal sig generates a sine wave and its first 999 upper harmonics with a value inversely proportional to the number of harmonic. Note the great compactness of this linguistic representation.

```
1  var sig, arr ;
2  arr = Array.series(size: 1000, start: 1, step: 1) ;
3  sig = Signal.sineFill(1024, 1.0/arr) ;
```

----

[3] Here phase is not taken into account, but the same principles apply.

A square wave is a sawtooth wave without even harmonics. The code is shown in the following example.

```
1  var sig, arr, arr1, arr2 ;
2  arr1 = Array.series(size: 500, start: 1, step: 2) ;
3  arr1 = 1.0/arr1 ;
4  arr2 = Array.fill(500, {0}) ;
5  arr = [arr1, arr2].flop.flat ;
6  // arr = [arr1, arr2].lace(1000) ;

8  sig = Signal.sineFill(1024, arr) ;
9  sig.plot ;
```

The amplitudes of the even harmonics must be equal to 0, the odd ones inversely proportional to their number.

The array arr1 is the array of amplitudes of odd harmonics. Note that step: 2 and arr1 are already properly scaled (3). The array arr2 (4) is created by the method fill that fills an array of the desired size (here 500) evaluating for each place the passed function. As we need an array composed of zeros, the function should simply return 0. Line 5 creates the new array arr, as it is more interesting, since it makes use of the methods flop and flat. See the example from the post window:

```
 1  a = Array.fill(10, 0) ;

 3  [ 0, 0, 0, 0, 0, 0, 0, 0, 0, 0 ]

 5  b = Array.fill(10, 1) ;

 7  [ 1, 1, 1, 1, 1, 1, 1, 1, 1, 1 ]

 9  c = [a,b] ;

11  [ [ 0, 0, 0, 0, 0, 0, 0, 0, 0, 0 ],
12    [ 1, 1, 1, 1, 1, 1, 1, 1, 1, 1 ] ]

14  c = c.flop ;

16  [ [ 0, 1 ], [ 0, 1 ], [ 0, 1 ], [ 0, 1 ], [ 0, 1 ], [ 0, 1 ], [ 0, 1 ],
17    [ 0, 1 ], [ 0, 1 ], [ 0, 1 ] ]

19  c = c.flat ;

21  [ 0, 1, 0, 1, 0, 1, 0, 1, 0, 1, 0, 1, 0, 1, 0, 1, 0, 1, 0, 1 ]
```

After creating two arrays of 10 places a and b (1, 7), a third array c is instantiated containing the two array a and b as its elements (9, as seen in 11 and 12). The method flop (14) "interlaces" pairs of elements from the two arrays (see 16). The method flat (18) "flatten" an array "eliminating all brackets": the structure of sub-array elements is lost. Compared to line 9, the result is an alternation of elements from the first and the second array (from a and b). In the example of the square wave, the result of line 5 is an alternation of elements from arr1 and zeros taken from arr2). As typical, the operation is actually already implemented in SC by a dedicated method, lace (6, commented): lace(1000) returns an array of size 1000 polling alternately arr1 and arr2.

4.5 Other signals and other algorithms

Other periodic signals can be generated in a trigonometric/procedural form, through a sum of sinusoids calibrated in amplitude. This is the case for example of the triangular wave. Even if described in a discrete form, such a method has its origin in the continuous domain. However, in the discrete domain other approaches are available, that are typically defined as "non-standard". For example, a geometric approach can be designed. The period of a triangular wave can be seen as consisting of four segments: the first in the range of $[0.0, 1.0]$, the second in $[1.0, 0.0]$, the third in $[0.0, -1.0]$ and the fourth in $[-1.0, 0]$.

```
1  // Triangular waveform by adding segments

3  var first, second, third, fourth, total ;
4  var size = 50 , step;

6  step = 1.0/size ;
7  first = Signal.series(size, 0, step) ;
8  second = (first+step).reverse ;
9  third = second-1 ;
10 fourth = first-1 ;
11 total = (first++second++third++fourth) ;

13 total.plot;
```

In the example, the variable size, the size of the array, contains the four segments, while step is the increase in the amplitude value. The first segment is then an array of type Signal filled by a number $step$ of values increasing by $step$: it contains values from 0 to $1 - step$ (7). The second segment follows the opposite path: the method reverse returns an array by reading backward the array on which it is called. Before, a step is added to each element of the array: second contains values from 1 to $0 + step$. The other two segments are obtained by generating two arrays third and fourth that subtract 1 respectively from second and first, that is, they are later "shifted down" (9, 10). Finally, the array total is obtained by concatenating the four segments. Note that the summing operations (as other operations on arrays) return an array in which each element results from the application of the sum to the relative element of the starting array. I.e.:

```
1  [1, 2, 3, 4]*2
2  [ 2, 4, 6, 8 ]
```

Such a method based on geometric synthesis allows one to obtain better results than a sum of sinusoids that can only approximate the triangular wave. Summing sinusoids would require infinite components to generate a triangular wave: the resulting edges are always "smoothed".

Again, the approach can also be discussed in relation to real/deferred time. The geometric approach must first calculate the segments in full, and then concatenate them: is not therefore a well-suited technique for real-time synthesis. An alternative approach is presented in the following example:

```
1  // Triangular waveform by series e sign change

3  var size = 100, freq = 2 ;
4  var step, val, start = 0, sign = 1 ;
5  var sig = Signal.newClear(size) ;
6  step = 1.0/size*(freq*4);
7  val = 0 ;
8  sig.size.do{ arg x;
9      if((val >= 1) || (val<=1.neg)) {
10          sign = sign.neg
11     } ;
12      val = val+(step*sign);
13      sig[x] = val ;
14 } ;

16 sig.plot ;
```

The idea, in this case, comes from the observation that the waveform is a straight line whose slope is reversed when a limit is reached, namely the two symmetric amplitude peaks, positive and negative ($[-1, +1]$). The variable step (6) determines the slope of the line. Four segments are in play, and freq multiplies them for each of their repetitions (in this case, 2). Let us consider two arrays of fixed size, both with triangle waves, yet the second is double the frequency of the first. It becomes clear that the slope of the second one is double that of the first ("it goes up and down at double speed") (Figure **4.8**).

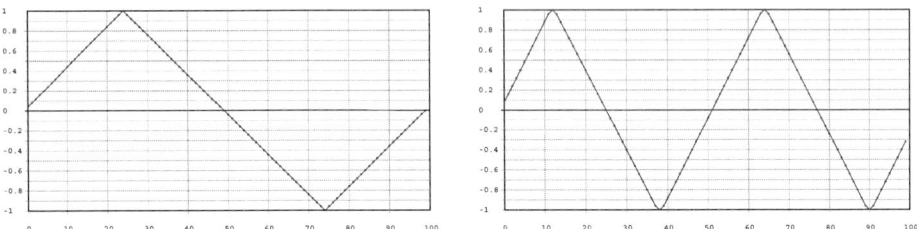

Figure 4.8 Triangular waves with frequency f and $2f$.

Once defined the meaning of step, we initialize val, the value of the first sample (7). The cycle iterates on the samples of sig (8-14). It checks that val is not greater than the positive peak or lesser that the negative, and in that case adds to val an amount step (12). The increase, however, is multiplied by sign.

```
1  0.216.sign
2  1

4  -0.216.sign
5  -1
```

Initially sign is 1, so that the increment value is unmodified. However, if a peak is exceeded, then sign is inverted by sign.neg (9-11). In other words, the wave raises up to 1 and then it decreases (a negative increase) up to -1.

Going back to the synthesis by sinusoidal components, the previously discussed example dealing with sum of sinusoids demonstrates the importance of the phase. In fact, the 50 phase-synced components increase together at the beginning of the cycle, producing the initial peak and then they interfere reciprocally until the negative peak phase π[4]. Figure **4.9**, top, shows the phases of the components (always 0) and the resulting signal.

---

[4] Note that pi and 2pi are two reserved words in SC.

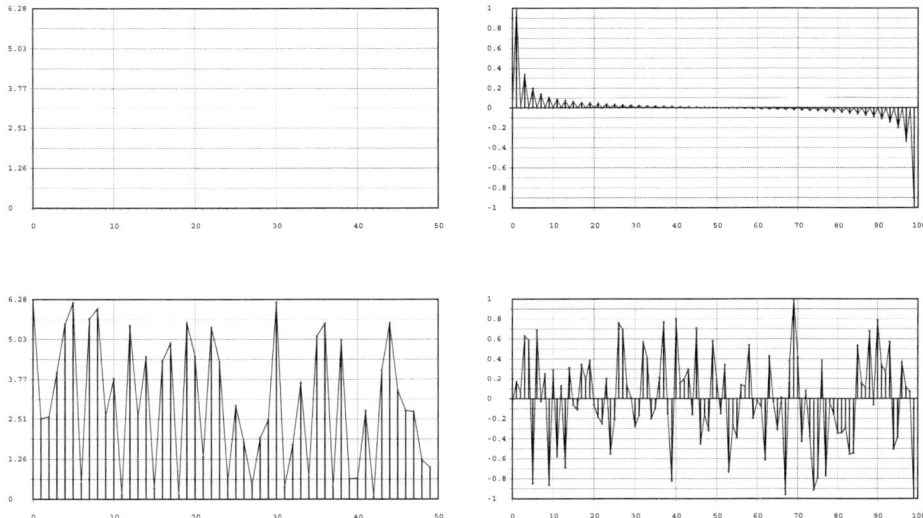

Figure 4.9 Sum of sinusoids: top, from left, stage 0 and signal; bottom, from left, and random phase signal.

In the following example the variable harmonics contains the number of harmonics. The array phases is instead created by the method fill (3) that requires us to specify a size (here harmonics) and a function that is applied to the calculation of each of the elements in the array. The function is {2pi.rand}, which results in a random value between $[0, 2\pi]$, i.e. a random phase in the whole cycle. The amplitudes of the components are all equal to 1 (4) and stored in amplitudes. Finally, sig is generated as a sum of harmonic sinusoids with the same amplitude but random phase.

```
1  var sig, harmonics = 50 ;
2  //var phases = Array.fill(harmonics, {0});
3  var phases = Array.fill(harmonics, {2pi.rand});
4  var amplitudes = Array.fill(harmonics, {1}) ;
5  sig = Signal.sineFill(1024, amplitudes, phases) ;
6  sig.plot ;
```

In this case, the amplitude of each component is equal to 1 but if it were generated by a function {1.0.rand}, it will receive a pseudo-random value between [0.0, 1.0] (between absence and maximum normalized value). If the previous code is modified according to this indication, by running the code several times it would be possible to notice that the sound changes, as the relevance of the components depends on the function of the method fill. Also, by increasing or decreasing the value of partials, the signal respectively will be enriched or impoverished in relation to higher components. Line 7 makes it possible to view the array of amplitudes as a broken line combining the discrete values: with plot it is possible to specify the maximum and minimum values of the desired range ([0, 1]). This is, as will be discussed later, a form of additive synthesis. Figure **4.9**, below, shows the phases of the components (random in the range [0, 2π]) and the resulting signal. Note that the harmonic content is the same but the waveform is radically different.

The introduction of the pseudo-random numbers enables us to create non-periodic signals. White noise is a signal whose behavior is predictable only statistically. This behavior results in a uniform energy distribution over the whole spectrum of the signal. White noise can be described as a totally aperiodic variation in time: in other words, the value of each sample is entirely independent from the previous and the following ones[5]. Therefore, the value of a sample x is independent of the value of the previous sample $x - 1$: x can have any value, always, of course, within the finite space of digital representation of the amplitude. Intuitively, the generation algorithm is very simple. In pseudo-code:

```
For every x in A:
      y = a*rand(-1, 1)
      A[x] = y
```

In the following example, the method fill (6) evaluates for each sample x a function that returns a random value within the normalized range [−1, 1] (rrand(-1.0,1.0)). The code specifies a sampling frequency (sr) and a duration in seconds (dur). The result (one second of white noise) is scaled to amp, and then, given a value of 0.75 for amp, the result is a (pseudo-) random oscillation in the range [−0.75, 0.75]. Notice how the function (by the very definition of $f[x]$) is actually calculated independently from x: being totally aperiodic, his behavior does not depend on time (that is, on x).

---

[5] It is said, therefore, that the signal has autocorrelation $= 0$.

```
1  var sig, amp = 0.75, dur = 1, sr = 44100 ;
2  sig = Signal.fill(dur*sr, { amp*rrand(-1.0,1.0) }) ;
3  sig.plot(minval:-1, maxval:1) ;
```

Softwares typically represent audio signals through a continuous curve that connects the values of amplitude. In the case of noise, the continuous representation is decidedly less clear than that of a discrete representation (Figure **4.10**), as here the image of a cloud of pseudo-randomly distribute points clearly emerges.

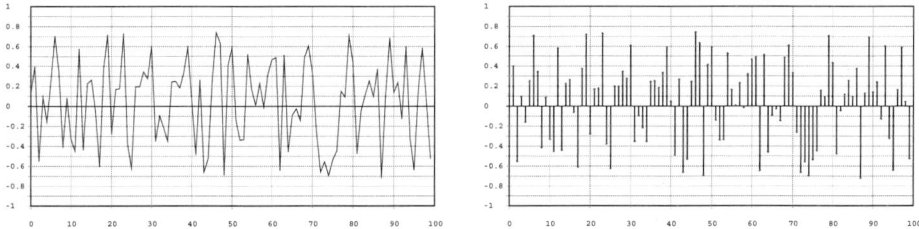

Figure 4.10   White noise: continuous curve and value dispersion.

However obvious it may be, it is worth remembering that a digital signal is in fact simply a sequence of values, on which it is possible to perform mathematical operations. Digital Signal Processing (DSP) techniques originate in this fact.

The next two examples are inspired by Miller Puckette[6]. A periodic function, such as the usual sine function, oscillates periodically between $[-1, 1]$ in normalized range. By calculating its absolute value, the negative part simply flips on the positive axis: by looking at the curve, it is apparent that the two semicircles are identical, thus resulting in a signal with twice the frequency of the original one. In general the application of the function of the absolute value is a rapid technique to make a signal jump by an an octave. Since the curve has only positive values (between $[0, 1]$ in the case of normalized values), it should

---

[6] M. Puckette, *The Theory and Technique of Electronic Music*, World Scientific Publishing Company, 2007, available on line:
http://msp.ucsd.edu/techniques.htm.

be translated to avoid an offset from zero (thus, the signal goes back from unipolar to bipolar): decreasing by 0.5 the signal oscillates in the range $[-0.5, 0.5]$ and it is again symmetrical between positive and negative (Figure in [abs]). In the following example, the method abs invoked on an object of type Signal returns another Signal in which each element is the result of the absolute value of the relative element from the starting signal (it is like calling abs for each element). The same also applies for subtraction and for multiplication: they are applied to each element of the array. The sine wave is then multiplied by amp and shifted "down" (towards the negative semiplane) of $\frac{amp}{2}$, here implemented with amp * 0.5[7]. The situation is represented in Figure **4.11**.

```
1  var sig, amp = 0.75 ;
2  sig = (Signal.sineFill(100, [1]).abs)*amp-(amp*0.5) ;
3  sig.plot(minval:-1, maxval: 1) ;
```

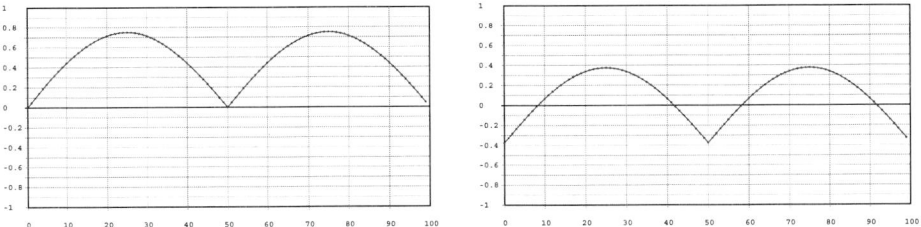

Figure 4.11   Absolute value of the sine function and translation ($A = 0.75$).

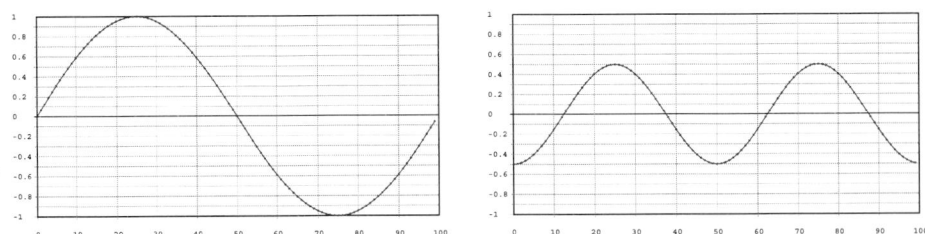

Figure 4.12 $y = sin(2\pi \times x)^2, y = sin(2\pi \times x)^2 - 0.5$.

---

[7] This is a typical practice in computer science. The multiplication costs, in computational terms, much less than the division.

A feature of the application of the absolute value is the generation of a signal which is asymmetrical with respect to the amplitude, and the presence of a very sharp edge (the flipping point) that introduces many higher components (higher harmonics). This feature can be avoided if, rather than absolute value, a power of two operation is applied: in fact the square of a negative value in the signal is positive, thus the resulting curve is substantially similar to what happens with the absolute value (with frequency doubling), but smoother (Figure **4.12**).

```
1  var sig, amp = 0.75 ;
2  sig = (Signal.sineFill(100, [1]).squared)*amp-(amp*0.5) ;
3  sig.plot(minval:-1, maxval: 1) ;
```

As a further example of minimal signal processing we can consider "clipping": which means all values above a certain threshold s (or lower than its negative) are reset to $\pm s$. Clipping also known as "digital distortion" because the same process occurs when a digitalized signal has an amplitude greater than quantization, and is therefore "cut" at its extremities. Clipping is a kind of radical limiter (as it limits the signal amplitude) and can be used as a distortion effect. SC provides the method clip2(t) that "cuts" a value out of the excursion $[-t, t]$ for $\pm t$. Consider the following example, where $t = 3$.

```
1  1.clip2(3)
2  1

4  -1.clip2(3)
5  -1

7  4.clip2(3)
8  3

10  -4.clip2(3)
11  -3
```

Although clip2 is already defined for Signal, to implement a "clipper" module is an interesting exercise.

```
 1  var sig, sig2, sig3, clipFunc ;
 2  sig = Signal.sineFill(100, [1]) ;

 4  clipFunc = { arg signal, threshold = 0.5 ;
 5      var clipSig = Signal.newClear(signal.size) ;
 6      signal.do({    arg item, index;
 7          var val ;
 8          val = if (item.abs < threshold, { item.abs },
 9              { threshold}) ;
10          val = val*item.sign ;
11          clipSig.put(index, val) ;
12      }) ;
13      clipSig ;
14  } ;

16  sig2 = clipFunc.value( sig ) ;
17  sig3 = clipFunc.value( sig, threshold: 0.75 ) ;

19  [sig, sig2, sig3].flop.flat.plot
20      (minval:-1, maxval: 1, numChannels:3) ;
```

As discussed, a function performs a behavior when it receives the message value with input arguments. The example defines a function that implements clipping, clipFunc: the required arguments are a signal (an object of type Signal) and a threshold value (threshold). The function clipFunc returns another Signal. Line 5 declares the variable that is assigned to the signal and immediately assigns an object Signal with the same size of the incoming signal, but filled with 0. The idea is to evaluate each of the elements (which represent audio samples) of Signal by cycling on it (6). In the loop, the variable val is the value to be written in the array clipSig. The value to be assigned to val depends on a condition.

If the sample is within the range $[-threshold, threshold]$ then there will be no clipping: val has the same value as the value of incoming sample. If the input value falls outside the range, then val will get the value of the threshold. The conditional is contained in lines 8-9. The evaluation is done on the absolute value of item (item.abs). It occurs in the interval $[0, 1]$. The output has the same absolute value of item or threshold. Thus, if the input value is negative, it would return a positive value. But the sign of the input value is retrieved in line 10 in which the value val is multiplied by the sign of item: in fact, item.sign returns ±1.

If item = -0.4:

- item.abs = 0.4
- is it lower than threshold = 0.5? Yes, then val = item.abs = 0.4
- item.sign = -1 (item = -0.4: it is negative)
- val = 0.4 * -1 = -0.4

Here are two examples, one that uses the default value of threshold (0.5), the other in which threshold is 0.75. Clipping tends to "square" the waveform, and, in fact, the resulting signal tends to sound like a square wave. Lines 19-20 are also worth discussing. The method plot is defined for an array, but one of its arguments is numChannels, that treats the actual sequence of values in the array as if it were interlaced values of numChannels channels. Then, with numChannels = 3 three curves are plotted, by considering each subsequent three-sample block as representing the values of the first, second and third channel. So, in our approach we need to create an array that includes the elements of the three arrays in the required order. With the method flop, an array of arrays is reshaped exactly in this way: in our case, three arrays of 100 values are reorganized into 100 arrays of three values. The method flat flattens the array, that at this point can be plotted by specifying numChannels: 3. The reader can try adding the appropriate postln messages to understand the example in more detail. The result is in Figure **4.13**.

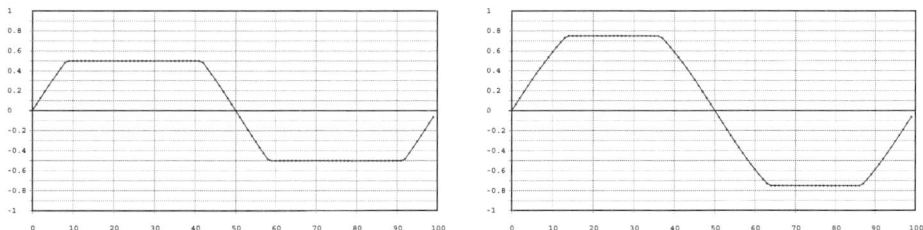

Figure 4.13 Clipping a sinusoid, threshold = 0.5, threshold = 0.75.

There are other, much more elegant ways to implement clipFunc. The next example uses the method collect, which, along with other similar methods, is inherited from the class Collection. These methods, once called on an collection-type object, return a new collection by applying a function for each element (each method implementing specific features). If the previous approach was typical of the imperative paradigm, these methods exhibit a rather functional attitude (and are also clear and concise).

In the following example, `collect` applies to each element its passed function.

```
1  clipFunc = { arg signal, threshold = 0.5 ;
2      signal.collect({ arg item;
3          var val, sign ;
4          sign = item.sign ;
5          val = if (item.abs < threshold, { item.abs },
6              { threshold}) ;
7          val*sign ;
8      }) ;
9  } ;
```

A function returns the value of its last expression. Here there is only one expression, which returns a `Signal` by applying a function to each element (similar to the previous implementation). The implementation aims at maintaining the modularity of `clipFunc`. It would have also been possibile to directly call `collect` on the `Signal` to be processed. The modular approach allows to re-write the previous cases related to the absolute value and the square functions. In particular, we can explicitly define two *transfer functions*:

$$W_{abs} : f(x) = x^2$$

$$W_{square} : f(x) = |x|$$

These functions behave as true processing modules of the input signal. A modular version of the absolute value function, written in a very compact fashion (note the use of |) could be written as:

```
1  absFunc = {|sig| sig.collect{|i| i.abs} - (sig.peak*0.5) } ;
```

In the code, the only thing of importance is the automatic elimination of the offset. The method `peak` returns the element of `Signal` that has the highest absolute value (the maximum amplitude of the signal). Assuming that the signal is symmetrical with respect to 0, each new value resulting from the application of the function is shifted by $peak/2$. If, for example, the signal's range is between $[-0.7, 0.7]$ then $peak = 0.7$: the processed signal will be included in $[0.0, 0.7]$ and is offset by $0.7/2 = 0.35$, so that the amplitude range is symmetrical around 0, in the range $[-0.35, 0.35]$.

4.6 Still on signal processing

The numerical nature of the signal allows us to define operations similar to the previous example on pre-existing material. The class Soundfile allows us not only to write to audio files, but to also access those available elsewhere (on a hard disk). The following code fills the object Signal sig with the contents of the audio file sFile through the method readData of SoundFile. The audio file is one of the examples provided with SC and can be accessed using Platform.resourceDir, a class that allows us to manage file paths regardless of the actual platform in use (OSX, Windows, Linux). Note how the size of sig has been defined relative to the number of samples of sFile by referring to its attribute numFrames.

```
1  var sFile, sig ;
2  sFile = SoundFile.new;
3  sFile.openRead(Platform.resourceDir +/+ "sounds/a11wlk01.wav");
4  sig = Signal.newClear(sFile.numFrames) ;
5  sFile.readData(sig) ;
6  sFile.close;
7  sig.plot ;
8  sig.plot ;
```

The following operation exploits the nature of the numerical sequence of the audio signal to implement a kind of granulation (to be discussed). In essence, the imported audio signal is divided into a number of blocks, known as chunks numChunks, each of which comprises a number step of samples (lines 12 & 13, here 500). The chunks are then randomly shuffled and reassembled. The implementation, indices (indexes) on line 17, is the sequence of the chunk indices, in this case starting from 0. This linear progression $(1, 2, 3, 4...)$ is shuffled through the method scramble (becoming for example, $9, 1, 3, 7...$). Then, by looping on each index, the corresponding chunk is retrieved from the original source (signal) and concatenated in sequence. The method copyRange copies from an array a set of elements specified by the arguments representing the start and end indices. It is likely that step is not an integer divider of the input signal. The remaining part (tail) is appended to the end of the new signal newSig.

```
 1 var sFile, sig, newSig ;
 2 var numChunks, step, rest, indices ;
 3 var block, tail ;

 5 sFile = SoundFile.new;
 6 sFile.openRead
 7    (Platform.resourceDir +/+ "sounds/a11wlk01-44_1.aiff");
 8 sig = Signal.newClear(sFile.numFrames) ;
 9 sFile.readData(sig) ;
10 sFile.close;

12 step = 500 ; // try with: 10, 200, 1000
13 numChunks = (sig.size/step).asInteger ;
14 // the tail...
15 tail = (sig.size-(step*numChunks)) ;

17 indices = Array.series(numChunks).scramble;

19 newSig = Signal.new;
20 indices.do({arg item;
21    block = sig.copyRange(item*step, (item+1)*step-1) ;
22    newSig = newSig.addAll(block) ;

24 }) ;

26 tail = sig.copyRange(sig.size-tail, sig.size) ;
27 newSig = newSig.addAll(tail) ;
```

This form of granulation of the signal allows to introduce a synthesis technique, that may be defined as "permutation synthesis". By "scrambling" the signal, it produces a new signal that more or less is related to the original signal. The lower the step the greater the recombination of the source signal. When such an operation is performed on a sinusoidal signal, the relation between the decrease of step and the increase of noise is apparent.

The following example implements a form of permutation synthesis. The process is similar to the previous example, with one relevant difference. Here the sinusoid is created by concatenating several copies (100, that is, 44100/period) of the signal sig (9-11). The signal is always divided into blocks of size step. In place of the scramble operation, the following permutation is implemented:

$$[\,0, 1, 2, 3, 4, 5\,] \rightarrow [\,1, 0, 3, 2, 5, 4\,]$$

Essentially, pairs of blocks are permuted. The permutation of the indices is calculated in two steps. First we create a series of even and odd indices, with a step equal to 2, starting from different values (1 and 0) (18-19). Then, the series are interlaced via flop and the final array is obtained by using flat (20). Since the permutation is periodic, the resulting signal has a very rich spectrum but has a periodicity that depends on both the frequency of the input sine wave and the size of step.

```
1   /* Distorsion by permutation on a sinusoid */

3   var sig, sig2, newSig ;
4   var numChunks, step, rest, indices ;
5   var block, tail ;
6   var period = 441;

8   // creation of the sinusoide
9   sig = Signal.new ;
10  sig2 = Signal.sineFill(period, [1]) ;
11  (44100/period).do(sig = sig.addAll(sig2)) ;

13  step = 50 ; // try with other steps
14  numChunks = (sig.size/step).asInteger ;
15  tail = (sig.size-(step*numChunks)) ;

17  // creation of the index sequence
18  a = Array.series((numChunks/2).asInteger, 1,2) ;
19  b = Array.series((numChunks/2).asInteger, 0,2) ;
20  indices = [a,b].flop.flat ;

22  newSig = Signal.new;
23  indices.do({ arg item;
24      block = sig.copyRange(item*step, (item+1)*step-1) ;
25      newSig = newSig.addAll(block) ;
26  }) ;

28  tail = sig.copyRange(sig.size-tail, sig.size) ;
29  newSig = newSig.addAll(tail) ;
```

The frequency becomes evident, not only by listening, but also by looking at the sine wave, distorted by permutation (Figure **4.14**).

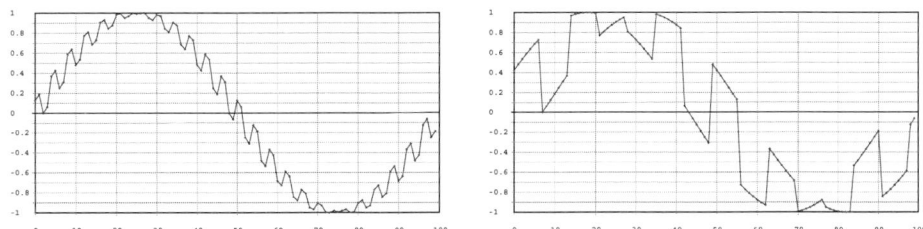

Figure 4.14 Sinusoid of 100 points and scrambling with step = 3 and = 7.

4.7 Control signals

A sinusoidal signal, as all perfectly periodic signals, completely lacks those dynamic features usually ascribed to a "natural" (or better, "acoustically interesting") sound: it is a sound, as Pierre Schaeffer said, without "temporal form", "homogeneous". However, sounds like this furnish the soundscape of mechanical and electrical modernity in the form of humming from street lamps, buzzing produced by fans, electrical impedance, engines, and much more. Apart from these cases, the temporal "shape" of a sound typically takes the form of a dynamic profile, a variation of the dynamics of the sound that is acoustically described in the form of an envelope curve. The phases of this curve are described using the terms attack/decay/sustain/release: hence the acronym ADSR. The simplest way to represent such an envelope is to use a broken line (Figure **4.15**).

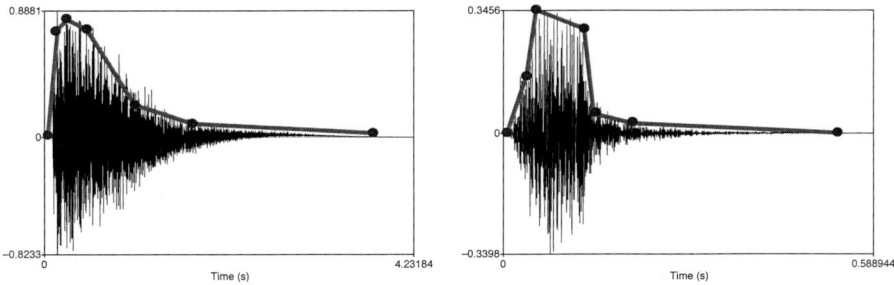

Figure 4.15 Description of the dynamic profile through a broken line (envelope).

A look at the envelope shows that it is just another curve, that is, another signal, differing from the signals so far considered by two important features:

1. it is not periodic (it consists of a single cycle). Therefore, the period of the envelope signal is equal to the duration of the audio signal: if the signal lasts for 2 seconds (that is, the period of the envelope) then the envelope has a frequency $1/2 = 0.5$ Hz. Note that the frequency falls outside the audible domain;

2. it is unipolar: assuming that the audio signal is normalized in the excursion $[-1, 1]$ (bipolarity), the envelope is included in $[0, 1]$ (unipolarity)[8].

Moving from analysis to synthesis, the task will be to reproduce the properties of the envelope (its "form") and to apply this form to an audio signal. The envelope is a typical *control signal*: a signal modifying - controlling - an audio signal. Being a signal, in order to represent an envelope we can still use an array. For example, a typical ADSR envelope could be described by means of the array of ten points in Figure **4.16**.

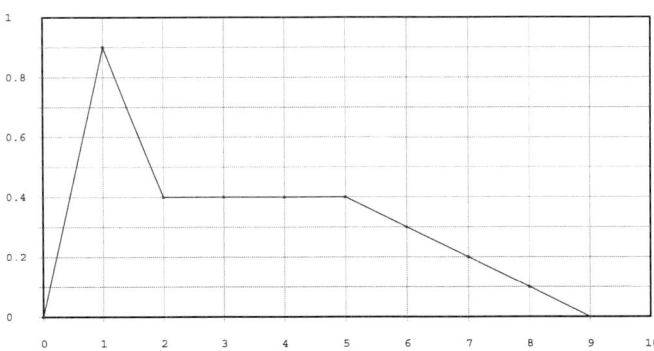

Figure 4.16   Array [0, 0.9, 0.4, 0.4, 0.4, 0.4, 0.3, 0.2, 0.1, 0].

In an interactive fashion:

---

8 Remember the case of the absolute value function that turns a bipolar signal into a unipolar one.

```
1  [0, 0.9, 0.4, 0.4, 0.4, 0.4, 0.3, 0.2, 0.1, 0].plot
```

In order to increase or decrease the amplitude of a signal we now understand that we can simply multiply the signal by a constant: each sample is thus multiplied by the constant. For example, if the constant $Amp = 0.5$, the signal amplitude is reduced by half. One could think of a similar constant in terms of a signal: as an array, with a size equal to that of the scaled signal, in which each element is the same value. Each value in the array to scale is multiplied by the respective value of the array Amp (always the same). The idea ceases to be an unnecessary complication at the moment in which the array Amp no longer always contains the same value, but instead contains variable values that represent precisely an amplitude envelope. Therefore, each sample of the audio signal (each element of the array) is multiplied by the related value of the envelope signal. The situation is represented in Figure **4.17**.

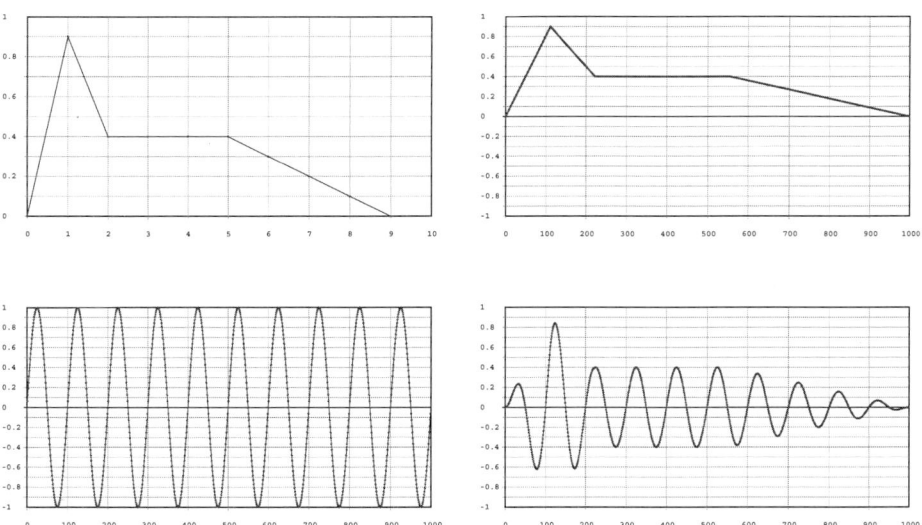

Figure 4.17   Envelope, resulting audio signal, audio signal, enveloped audio signal.

To fully represent the duration of an audio signal, the size (length) of the envelope must be the same of the audio signal. Here a fundamental point shows up, to be discussed again: typically, a much smaller number of points is enough to represent an envelope with respect to an audio signal, that is assuming 1 second audio signal of CD-quality, and the adequateness of the envelope in Figure **4.17**, the ratio between the two could be something like $10/44, 100$. This is in fact a relevant feature that identifies a control signal.

Moving on to the implementation in SC (below), we can generate a sinusoid of a duration equal to a second by means of the method waveFill. We have chosen a low frequency, only for sake of visibility. The envelope array to be implemented is made up of ten segments: since these are to be distributed throughout the duration of the audio signal, each of the segments will include $44, 100/109$ samples. This value is assigned to the variable step.

The task now is to generate the arrays that describe the four ADSR segments (att, dec, sus, rel, 8-11), and then concatenate them in a unique array env. Since they must be multiplied by an object Signal then they must also be of the Signal type. The method series(size, start, step) creates a linear progression of size values from start with an increase of tt step.

The attack starts at 0 and reaches 0.9, it occupies a segment, i.e, 4410 samples. The value 0.9 should thus be reached in 4410 points: the increase of each will therefore be 0.9/4410 - in other words, 0.9/step. The last point will receive a value of $0.9/step \times step = 0.9$. In relation to dec the value to be reached is 0.5, along the duration of another segment. In this case the progression starts at 0.9 and the increment is negative: it must go down by 0.5 in one step, therefore the increase is -0.5/step. In the third case (sus), the value is constant at 0.5 for a duration of 4 steps, then the increase is 0. The case of rel is similar to that of dec. The new envelope signal env (13) is obtained by concatenating the four segments where it is used as a multiplier for sig, in order to get the enveloped signal envSig (14). Line 16 draws the signal, the envelope and the enveloped signal.

```
1 var sig, freq = 440, size = 44100, step ;
2 var env, att, dec, sus, rel, envSig ;

4 step = 44100/10 ;
5 sig = Signal.newClear(size) ;
6 sig.waveFill({ arg x, i; sin(x) }, 0, 2pi*50) ;

8 att = Signal.series(step, 0, 0.9/step) ;
9 dec = Signal.series(step, 0.9, -0.5/step) ;
10 sus = Signal.series(step*4, 0.4, 0) ;
11 rel = Signal.series(step*4, 0.4, -0.4/( step*4)) ;

13 env = att++dec++sus++rel ;
14 envSig = sig * env ;

16 [sig, env, envSig].flop.flat.plot(minval:-1, maxval: 1, numChannels: 3) ;
```

The above method is quite cumbersome, and quite theoretical in SuperCollider. SC provides a class, Env specializing in building envelopes. Env assumes that an envelope is a broken line that connects amplitude values over time and provides various interpolation methods for intermediate values. Consider the following two arrays v and d:

```
1 v = [0,   1,    0.3,    0.8,    0] ;
2 d = [  2,   3,    1,    4  ] ;
```

A similar pair is typically used to specify an envelope:

- v: indicates the points that define where the envelope steepness change (i.e. all peaks and valleys);
- d: indicates the duration of each segment connecting two points.

Therefore, the array of the durations d always has a size equal to the size of amplitude v less 1. In fact, d[0] (= 2) indicates that the duration between v[0] (= 0) and v[1] (= 1) is 2 units of time (in SC, they are seconds: but in reality they can be conceived as abstract units, as we will see). Through the two arrays v and d it is thus possible to describe a time (d) profile (d). In the example, however, it is still not specified what happens for each sample comprised between 0 and 1. The way in which the intermediate samples are calculated depends on the interpolation mode. In the following example e1, e2 and e3 are instances of Env, using the same pair of arrays v, d, but with different modes of interpolation (linear, discrete, exponential). The final example, e4, shows a different possibility: an array of interpolation modes, one for each segment (the last mode is 'sine').

```
1  var v, d, e1, e2, e3, e4 ;

3  v  = [0,    1,     0.3,      0.8,      0] ;
4  d  = [   2,    3,         1,      4   ] ;

6  e1 = Env.new(v, d,'linear') ;
7  e2 = Env.new(v, d,'step') ;

9  v  = [0.0001,   1,      0.3,      0.8,     0] ;
10 e3 = Env.new(v, d,'exponential').asSignal ;
11 e4 = Env.new(v, d,[\linear , \step ,\exponential , \sine ]) ;

13 [e1, e2, e3, e4].collect{|i| i.asSignal}.flop.flat.plot(numChannels:4) ;
```

Note that in the case of an exponential envelope the starting value cannot be $= 0$: the first value of v is then redefined with a value close to zero. The meaning of the parameters is shown in Figure **4.18** commenting the output of the method plot. The class Env inherits directly from Object and is therefore not an array-like object. Typically Env is used to specify an envelope for the real-time usage (as we will see). However, when an object Env receives the message asSignal, it returns a Signal object containing a sampled envelope. The method is used (on line 13) to obtain arrays from envelopes, in order to apply the plotting technique already discussed.

Through asSignal the class Env allows one to use, even in deferred time, a specification for envelopes. In addition, the class includes some constructors that return particularly useful envelopes. Two examples:

- triangle takes two arguments: the first indicates the duration, the second the peak value of a triangular envelope (i.e. the peak at half the duration).

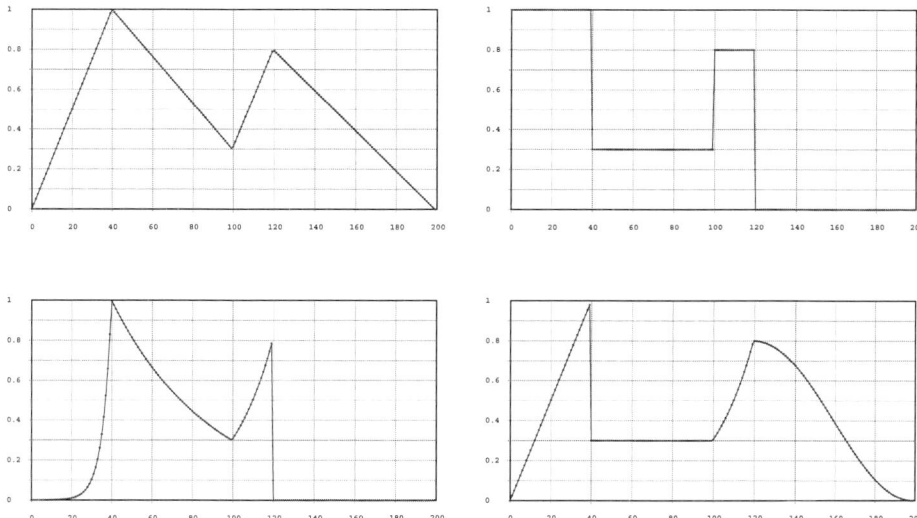

Figure 4.18   Env: interpolation types.

- perc: allows to define a percussive envelope (attack + release). The arguments are attack time, release time, peak value and value of curvature.

```
1  var sig, freq = 440, size = 1000 ;
2  var envT, envP ;

4  sig = Signal.newClear(size) ;
5  sig.waveFill({ arg x, i; sin(x) }, 0, 2pi*50) ;

7  envT = Env.triangle(1,1).asSignal(size);
8  envP = Env.perc(0.05, 1, 1, -4).asSignal(size) ;

10 [sig, envT, sig*envT, envP, sig*envP].flop.flat
11    .plot(minval:-1, maxval: 1, numChannels: 5) ;
```

The envelope types triangle and perc, and their applications using a sine wave, are represented in Figure **4.19**.

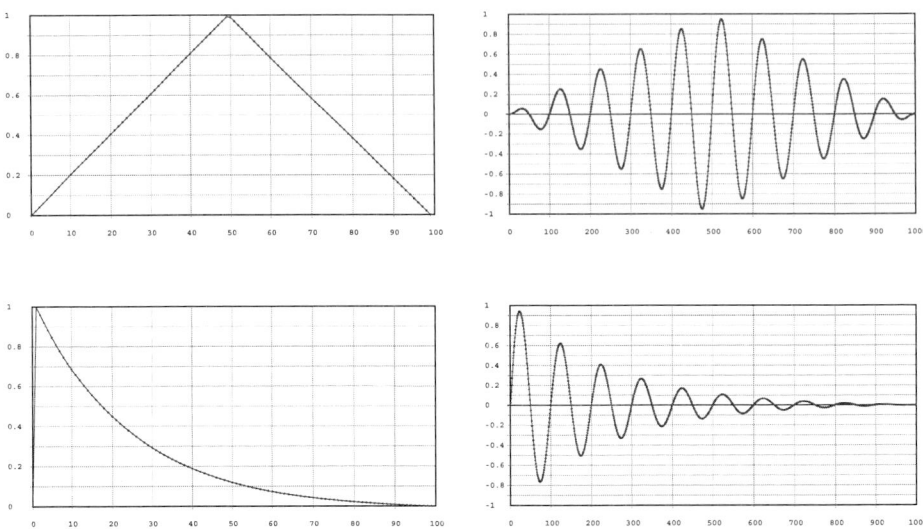

Figure 4.19   Envelopes with Env, triangle (top) and perc (bottom).

An envelope signal can also be applied to an audio signal of concrete origin (i.e. from an audio file). In the following example, the signal sig is obtained by importing the content of sFile, which has to be normalized so that its peak is equal to 1 (6). An envelope signal env is applied to sig: env is obtained through two arrays filled with pseudo-random numbers, v and d. The first varies in the range $[0.0, 1.0]$ (it is a unipolar signal). To avoid offset in amplitude, the first and the last value of the array are set to 0, and added respectively to the head and tail (11). The array d has a size equal to v -1, as required by the syntax of Env. Time intervals vary in the range $[0.0, 4.0]$ (12).

```
1  var env, v, d, breakPoints = 10 ;
2  var sig, sFile ;

4  sFile = SoundFile.new;
5  sFile.openRead(Platform.resourceDir+/+"sounds/a11wlk01-44_1.aiff");
6  sig = Signal.newClear(sFile.numFrames).normalize ;
7  sFile.readData(sig) ;
8  sFile.close;

10 v = Array.fill(breakPoints-2, { arg i ; 1.0.rand }) ;
11 v = v.add(0) ; v = [0].addAll(v) ;
12 d = Array.fill(breakPoints-1, { arg i; 4.0.rand }) ;

14 env = Env(v, d, 'lin').asSignal(sig.size) ;

16 [sig, env, sig*env].flop.flat.plot(minval:-1, maxval: 1, numChannels: 3) ;
```

Two pseudo-random envelopes generated by the same code are drawn in Figure **4.20**: on each code evaluation, the envelope takes on a different form, apart from the two extremes set to 0.

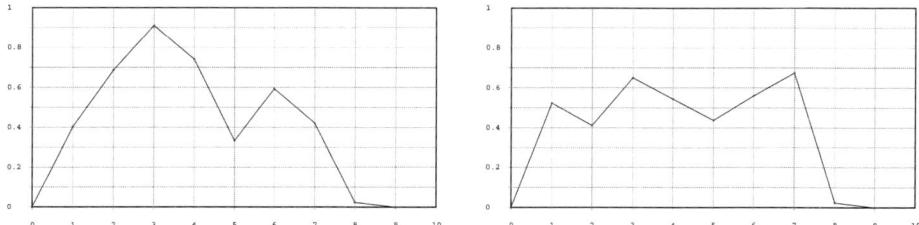

Figure 4.20   Pseudo-random envelopes.

4.8 Conclusions

The goal of the chapter was to introduce the concept of a digital signal through operations that are possible thanks to its numerical nature. We have also had the opportunity to see how it is possible to change a signal by means of another signal. In particular, a control signal is a signal that requires a temporal resolution much smaller than an audio signal, its frequency standing far below the audible frequency ("sub-audio range"). A control signal typically modifies an audio signal. However, the chapter was also intended as an opportunity to explore some aspects of the SuperCollider language. At this point, it is important to move along and expand these ideas by turning to the more typical *modus operandi* of SC, real-time sound generation.

5 SC architecture and the server

So far we have discussed many aspects of SuperCollider, however the synthesis and processing of audio, one of the most famous features (and strengths) of SC, has not as yet been examined. It is therefore time to address this issue, but, as usual, some patience maybe required in order to get to the bigger picture.

5.1 Client vs. Server

As we have observed, when installing the SC application we obtain two autonomous components, a server and a client. The first is called *scsynth*, the second *sclang*. The latter is the interpreter for the SuperCollider language that has been discussed in previous chapters. But it also works as a client for the audio server. In summary, the sound server is an audio provider, capable of generating and managing a large variety of signal processes. The server is an audio "provider", and its services must be requested by a "customer", technically a "client". This type of software organization is thus called "client/server architecture". Figure **5.1** describes a generic network architecture: multiple clients communicate with a server by exchanging messages.

The SC application is thus built upon a client/server architecture, that splits the computation into two modules, one for the request and the other for the supply of services. In SC the client and server communicate over a network (to be precise, according to the network protocol UDP) by means of messages written in a specific communication protocol (a "code" known to both), named Open Sound Control (OSC). OSC is commonly used in multimedia applications

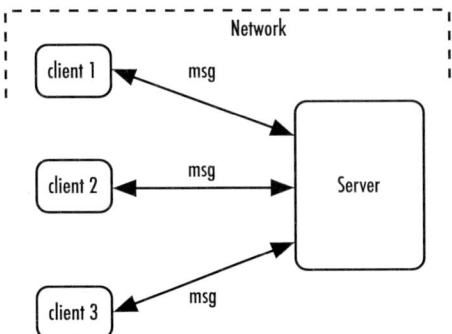

Figure 5.1 Generic client/server architecture.

(for example, it is implemented in Max/MSP, PD, EyesWeb, Processing, etc.)[1]. The OSC protocol, in itself, does not specify a semantics, that is, it does not define the meaning of a closed set of possible messages: rather, it defines a syntax for constructing messages that are compliant to the protocol. In this way, applications and users can define their own messages and associate them a specific semantics. In our case, scsynth defines a set of OSC messages that grant access to all its operations.

For the avoidance of doubt, the network we are speaking about is defined on an abstract level. This means that the client and server can be running on the same machine. It is what happens when you execute the SC *application*: without any extra settings, the interpreter is already configured to communicate over a network as a client for the audio server. Repetita iuvant. By opening SC, two programs are already configured to work together over a network, sclang and scsynth. The latter (the server) is an audio synthesis engine, powerful, efficient, low-level application, that is, not very "intelligent".

The interpreter –sclang– has two functions:

1. it is the client: in other words, it is the interface that allows the user to write and send OSC messages to the server. To write a letter to the server, you need to have a sheet of paper and a postman that delivers it: sclang does both.
2. it is the interpreter for the SC language: OSC messages may be quite cumbersome to write, and share with the server this low-level perspective. As we have abundantly seen, the SC language is on the contrary a high-level

---

[1] Historically, James McCartney, the first author of SC, has also contributed to the definition of the protocol itself.

language. The SC code, when it refers to the server, is then translated by the interpreter into the proper OSC messages that are sent to the server. The poem that you write in the SC language is paraphrased by the interpreter into the OSC prose that the more mundane server understands.

The situation is depicted in Figure **5.2**. Communication between the client- and the server side happens via OSC messages that the client sends to the server. The sclang interpreter sends OSC messages in two ways:

1. **directly**. In other words, the sclang interpreter is yet a good place for you if you want to talk to the server with the OSC code (that is, sclang is a place where to write directly OSC messages);

2. **indirectly**. The SuperCollider language (at a more abstract, "high" level) is automatically translated by the interpreter into OSC messages (at the server level) in order to be sent to scsynth (this is the so-called *language wrapping*).

To sum up, starting from the sclang interpreter you can write poetry by relying on the translation into prose performed by the interpreter (which acts as a translator and postman) or you can directly write in OSC prose (in this case, the interpreter is only the postman). One might ask why bother with such an architecture. The advantages are the following:

- **stability**: if the client experiences a crash, the server continues to run (i.e. the audio does not stop, and this is an important feature in a concert, installation, performance ...), and vice versa.
- **modularity**: one thing is the synthesis, another one is control. Keeping the two functions distinct allows to control scsynth even from applications that are not sclang: the important thing is that they know how to send the right OSC messages to the server. The server is democratic (everyone can get audio services) and bureaucratic at the same time (the important thing is to respect the OSC protocol and the semantics defined for scsynth). As shown in Figure **5.2**, other applications can send messages to the server[2].
- **remote control**: the network we are talking about can be either internal to the computer, or an external one. When dealing exclusively with audio syn-

---

[2] Over the years, various clients for scsynth have been developed in other programming languages such as Clojure, Java, Scala. In this way, experienced programmers of other languages can access scsynth without knowledge of the SC language.

thesis, typically the two components work on the same computer by sharing a local address (it is the default situation). But client and server may well be on the network and, with some *caveat*, even on the two opposite ends of the globe, communicating via internet.

The main disadvantages of such an architecture are two:

1. the circulation of messages introduces a small delay (which may be significant, however, if considering time sensitivity of the audio domain);
2. a great time density of the messages on the network can overloaded the latter, and message handling can cause a delay.

It should also be noted that it is very rare to incur similar problems.

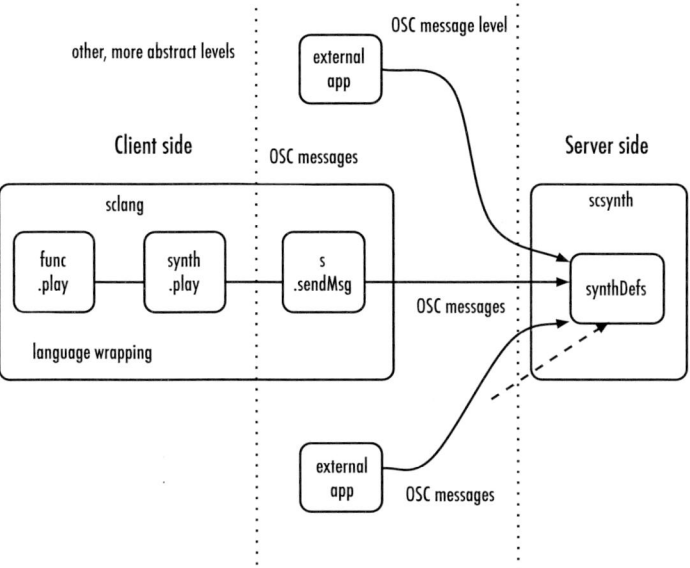

Figure 5.2   Client/server architecture in SC.

The above discussion makes it clear that sclang is only one of the possible clients for scsynth. However, sclang, being designed specifically to be used with scsynth, it is somehow the privileged client, because it provides the user an integrated, high-level language and translates it in OSC messages[3].

As mentioned, the server can be controlled (only) through OSC messages: that is, scsynth exposes an interface of OSC commands. However, the SuperCollider language defines a set of classes that represent all the objects in the server, and manage the relative OSC communication with it. So, what we learned previously in relation to the SC language can be fully exploited in the interaction with the server. Hence, the user must be aware that there is circulation of OSC messages between the client and the server. But, apart from special cases, this low-level business can be safely ignored.

5.2 Ontology of the server as an audio synthesis plant

So, scsynth is an audio synthesis engine that is programmable and controllable in real-time. It is not easy at first to be able to keep in mind the relationship between all the elements relevant to the server. In fact, there are some purely computer-science notions (the client/server architecture), other that are typical of the digital audio domain (sampling, quantization, etc.), other that are borrowed from computer music (the concept of "instrument", from Music-N languages), other that are rooted in electronic music *tout court* (many synthesis techniques are digital implementations of analogue methods, as well as the idea of "patching"), still other notions stem from analog audio even if typical also of the electroacoustic practice (the concept of buses in the mixer, that is also present –in a digital fashion– in DAWs, Digital Audio Workstations).

It is therefore convenient to introduce a metaphorical framework –a bit "quirky" indeed, but hopefully useful– and think about the server as a chemical plant for the synthesis of liquid materials. Such a framework allows to introduce the ontology of the server, that is, the set of elements that compose it and their mutual relations. Each element will then be addressed analytically, in the proper context of scsynth. The following discussion will start from Figure **5.3**.

---

[3] In particular, the element that makes sclang a privileged client is the opportune formatting of SynthDefs that it provides, so that SynthDefs can be sent to the server. This operation can be implemented into other clients, but it is objectively cumbersome. We will discuss SynthDefs in detail later.

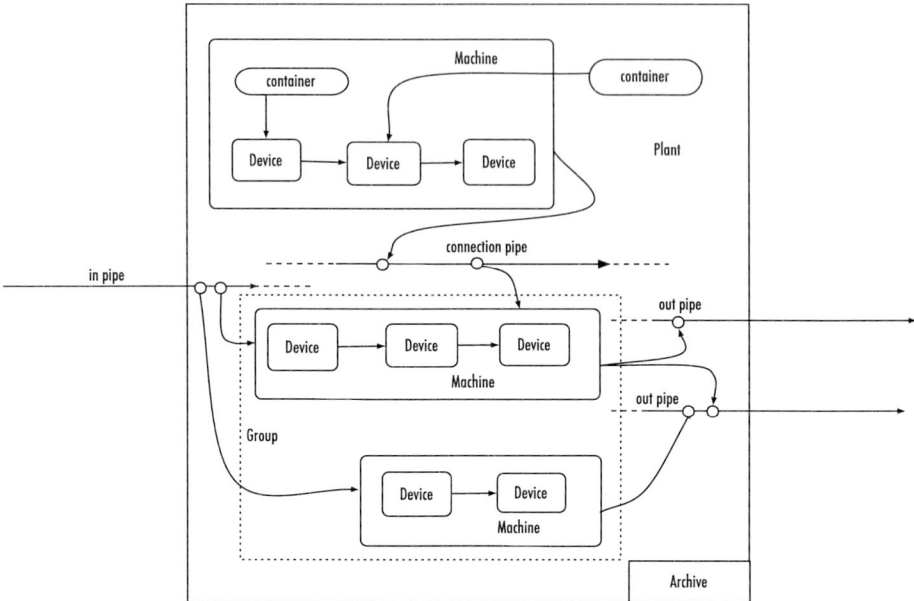

Figure 5.3   The audio server as chemical plant for the synthesis of liquids.

1. The whole production process is managed not by a machine but by a whole plant;
2. To synthesize a liquid, a complex machinery is required;
3. A machine consists of specialized devices in which liquids are being transformed. Devices are connected through inner pipes;
4. Machinery must be designed providing the relationships between device components. From a project schematics, any number of identical machines can be built. Projects are stored into the plant archive for future reuse;
5. Once built, the machine can not be changed in its internal structure;
6. But an employee can control its behavior from outside through levers and controls, as well as monitor its operation;
7. A plant may comprise more machines working in parallel;
8. Groups of autonomous machines can be coordinated;
9. When the plant is in operation, fluids flow through the pipes at constant rate, but can never stop;
10. The liquids can flow in the pipes at two (but always constant) different rates, and in particular at control and synthesis rates;
11. Liquids, however, can be stored in limited quantities in special containers, from which they can be drawn at need. These liquids do not flow, but, through specialized devices, they can be poured into a a flowing liquid;

12. Typically (although not necessarily) a machine is given a device equipped with a pipe that allows to let out the liquid outside. Other times, it can also have a device with an incoming pipe from which to receive a liquid that comes from other machinery;

13. The circulation of the liquids between the plant and the outside (e.g. water from the inlet, the synthesized product in output) or between different machines in the plant occurs through special pipes. The former are pipes devised for entry/exit, the second are connection ones. Through the latter, liquids can then circulate in the plant and are available to other machines. These pipes therefore allow you to connect different machines;

14. The connecting pipes disperse liquids that not exited through the outlet pipes of the plant in the waste plant. Liquids do not pollute and their dispersion is not relevant;

It is possible now to reconsider the previous points, by appropriately replacing the names in Figure **5.3** with those in Figure **5.4**.

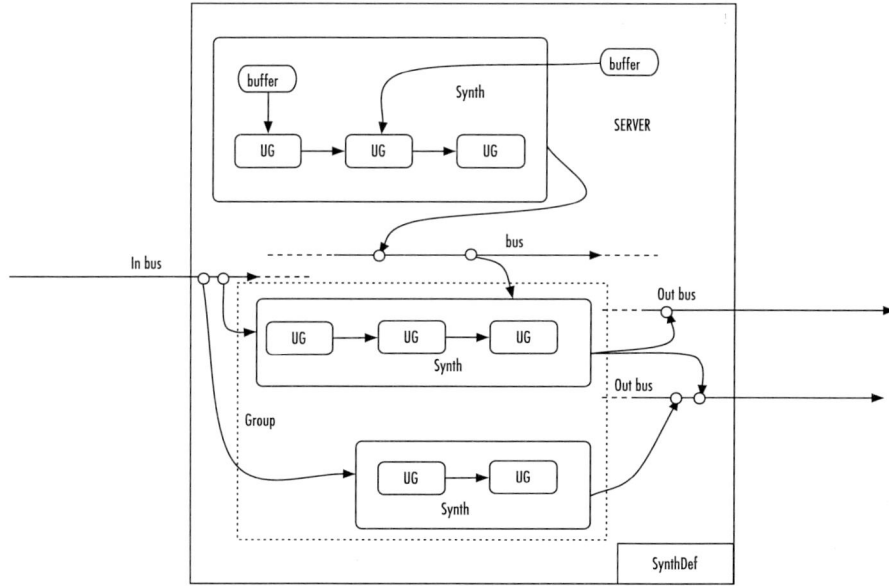

Figure 5.4 Components of the audio server scsynth.

Observation 1
The whole production process is managed not by a machine but by a whole plant

It is useful to consider the server not so much as a machine, rather as a real audio synthesis plant, an environment with a complex internal organization. The server must be set in motion ("booted"). Note that when you open the IDE, the interpreter is already active (below left, its relative GUI is green), but the server has not started yet. To start the server, you can use the GUI, that is, press the window and select boot or evaluate s.boot, because the environment variable s is associated by default in the interpreter to the sound server.

Observation 2
To synthesize a liquid, a complex machinery is required

In order to synthesize an audio signal, a software machine is needed. In SC, the latter is termed Synth: a "synth" is precisely an audio synthesizer. Here synthesizer is to be understood literally with a reference to an instrument, the analog synthesizer. A synth is an instrument that sounds.

Observation 3
A machine consists of specialized devices in which liquids are being transformed. Devices are connected through inner pipes

To generate an audio signal, a synth requires the user to specify which synthesis/processing algorithms for the audio signal have to be used. In SC, following the tradition of the *Music N* family of languages, synthesis/processing algorithms are implemented in "UGens" (→ *Unit Generator*): a UGen is simply a software device that processes or synthesizes audio signals. For example, SinOsc is a UGen that generates sinusoidal signals. The UGen is the very basic component of a synth, an atomic object that is an audio primitive in SC[4]. A synth is just a synthesizer, a synthesis device built with UGen components. A UGen has its analog counterpart in a hardware component that can be wired to other hardware components.

Observation 4
Machinery must be designed providing the relationships between device components. From a project schematics, any number of identical machines can be built. Projects are stored into the plant archive for future reuse

---

[4] The equivalent in Csound is the *opcode*, and the same applies to the generating units that correspond to a block in graphical languages such as Max/MSP or Pure Data.

The SC client, therefore, asks the server to build and operate one (or more) synth on its behalf. To meet the demand, the server has to know which pieces (UGens) to use and how to combine them (how to "patch" them into a "UGen graph", a graph of UGens). Since it is likely that the user may need several times the same devices, SC provides an extra step. The user first provides the server a definition of a synth (a "SynthDef"), a kind of detailed schematics of how the desired synth should be built, and then asks the server to effectively build a synth following that project. View from the perspective of analog hardware, a SynthDef is like the wiring diagram upon which to build a synthesizer. A SynthDef associates a name n to a configuration of UGens u, so that synths of type n can be built, that generate signals through the relationships defined for the UGens u. Once created, the SynthDef can be eventually stored on the hard disk and therefore remain always available to the user even after the SC application is closed. In other words, the user can create a library of SynthDefs (understood as projects or as templates from which to create synths) and, when appropriate, ask the server to create a synth from a certain SynthDef.

Essentially, to use SC as a synthesis engine, two steps at least have to be taken:

1. define a SynthDef (define the design of the synthesizer)
2. instantiate a synth from a SynthDef (build the synthesizer)

Observation 5
Once built, the machine can not be changed in its internal structure

A SynthDef is a diagram, a schema: it is a static object. Once sent to the server, it is immutable. If it has two inputs, other could not be added. On the other hand, it is always possible to send the server a new SynthDef implementing the desired changes, as well as to overwrite an existing SynthDef.

Observation 6
But an employee can control its behavior from outside through levers and controls, as well as monitor its operation

The project of a synthesizer is described in a SynthDef by means of a UGen graph. The latter may provide input arguments to be controlled by the user: these arguments are precisely parameters for the computation –carried out within the UGen graph– of the value (the amplitude of the signal, remember that a signal is a sequence of numbers) to be returned in output.

Observation 7

A plant may comprise more machines working in parallel

Each machinery in the hydraulic example represents a synth, or, musically speaking, an instrument -or even a voice. On the server a very large number of synth can reside, the default value is 1024, but it can be increased by the user (if computational resources are available). All synthesizers may operate in parallel.

Observation 8

Groups of autonomous machines can be coordinated

Synths can work in parallel, but it is also possible to control them as a single group. That is, it is possible to send coordinate communications to different synths if they are part of the same "group".

Observation 9

When the plant is in operation, fluids flow through the pipes at constant rate, but can never stop

We assumed that the liquid is the metaphoric counterpart of the audio signal. The choice of the liquid depends on the fact that at this point we are not considering signals *per se*, but signals in real-time. In real-time, signals are indeed sequences of amplitude values, but with the added constraint that these samples should be inexorably calculated at a uniform rate over time (typically, but not necessarily, in the case of audio signals, $44, 100$ times per second, the sampling rate of the audio CD). At each instant of time, a new sample must be computed: each synth performs all the calculations requested by all UGens that compose it and returns a value. In other words, at each instant, all the UGen graph must be traversed, regardless of its complexity. Warning: if you consider the hydraulic example, this means that if a drop enter from a pipe in the instant x, the instant later $(x + 1)$, it must have passed through the entire machine, and spit in output. That is: in each machine, the flowing of the liquid along the pipes that connect internally the devices is literally instantaneous.

Observation 10

The liquids can flow in the pipes at two (but always constant) different rates, and in particular at control and synthesis rates

So at each instant a new drop must come out by a machine after traversing the entire complex of devices. The devices do not necessarily, however, update their behavior on instant basis. In some cases, they may change their behavior once every n instants. A control signal is typically a signal that changes less over time of an audio signal and that therefore can be calculated at a lower time rate. For example, let us consider a signal that represents a vibrato, a frequency oscillation. This oscillation will vary, by way of example, to a maximum of 10 times per second: it is a sub-audio signal. Thinking of the curve representing such an oscillation, its frequency can be described as $f = 10$ Hz. The Nyquist theorem states that to represent a 10 Hz oscillation we need a sampling rate of just $10 \times 2 = 20$ Hz. In such a situation, a rate of $44, 100$ is a waste of computational resources, as we would update at audio rate a signal that changes at a much lower rate. If we assume to work at audio rate (e.g., $44, 100$ Hz), the value of the vibrato signal by which to multiply the audio signal can be computed at time 0, kept constant for n audio samples, then recomputed at time $n + 1$, so that the new value can be used to update the vibrato oscillation, again kept constant for n samples, and so on. A similar signal is not computed at *audio rate*, rather at *control rate*. As we will see, the UGen compute signals when receiving the messages ar or kr: the resulting signals will be updated respectively at audio- or control (*[k]ontrol*) rates. Note that we are talking about an update rate for signals, we are not considering an absolute number of samples. SC always generates an audio signal in real-time by traversing the UGen graph at audio rate: but some signals which intervene in the synthesis are updated at a lower rate control.

Observation 11

Liquids, however, can be stored in limited quantities in special containers, from which they can be drawn at need. These liquids do not flow, but, through specialized devices, they can be poured into a a flowing liquid

A buffer is a temporary memory that allows to store audio data required by a certain synthesis algorithms. For example, consider reading a sound file for playback. There are indeed UGens specialized in this operation. The content of the audio file must then be read from the hard disk and stored in a block of temporary memory (RAM). In SC a buffer is a block that the server allocates on the RAM on request, so that it can be used to store data. The audio signal stored in the buffer is in itself static (it is just a data sequence): and yet there may be a UGen that reads the contents from the buffer in real-time and sends it to the sound card. Looking at Figure **5.4**, it can be noted that there are two types of buffers. Some are internal to the synth, allocated directly when a synth is created and not accessible from outside. Others are allocated outside the synths and thus accessible to any synth.

Observation 12
Typically (although not necessarily) a machine is given a device equipped with a pipe that allows to let out the liquid outside. Other times, it can also have a device with an incoming pipe from which to receive a liquid that comes from other machinery

The synthesized digital signal must be sent to the sound card so that it converts it into an electrical signal and sends it to the speakers. Some specialized UGens are available to accomplish this task. They include inputs but not outputs: in fact, once the signal has passed through these UGens, it is no longer available for further processing in SC, as it is sent to the sound card. A typical example is the UGen Out. Typically, UGens of this type occupy the last places in a UGen graph representing a synth. If an output UGen is omitted, the signal is computed as required by other UGens in the UGen graph, but not sent out from the synth, e.g. to the sound card (a waste of mental and computational labor, and a classic mistake of the novice). Let us suppose then to connect an input device to the sound card, such as a microphone. A specialized UGen can make available to the synthesizer such a signal, so that it can be processed (e.g. distorted). To do this, SC features the UGen SoundIn.

Observation 13
The circulation of the liquids between the plant and the outside (e.g. water from the inlet, the synthesized product in output) or between different machines in the plant occurs through special pipes. The former are pipes devised for entry/exit, the second are connection ones. Through the latter, liquids can then circulate in the plant and are available to other machines. These pipes therefore allow to connect different machines

Thus, the server provides UGens for the communication with the sound card, in input (e.g from a microphone) and output (e.g. to speakers). These communication channels are called "buses", according to a term derived from the technology of mixer[5]. Even if the notion of "channel" could be close to that

---

[5] Just to be clear, the term has nothing to do with "bus" as a means of transport. Even if a bus serves for transport, any comparison would be misleading in this respect. A bus does not move anything in a proper sense. We will discuss buses later

of bus, it is probably better and less misleading to think about buses in terms of "pipes". In any case, the bus system must not be thought of as a system of pipes that statically connect various machines, rather as a system of available pipes to which machines can be connected. For example, all synths meant to process a signal that comes from the outside can be connected to the bus which is responsible for reading the input from sound card (the pipe where a liquid from outside flows). All synths sending their output signals to the sound card may be connected to the bus that handles that task: in that case, as a digital signal is just a sequence of numbers, the signals on the output bus simply add up. The buses so far mentioned are specialized for audio signals (they are *audio buses*), but there are also buses specifically dedicated to control signals (*control buses*). Buses are identified by a progressive number, an index starting from 0. For control buses, the indexing is progressive and there are no particular aspects to be considered. Audio buses require extra attention, as indexing requires to take into account also the special cases related to communication with the sound card. The first indices are dedicated to "external", "public" buses that handle the I/O communication with the sound card. The next indices are dedicated to the "internal", "private" buses[6]. What is the role of the internal buses? They are needed to connect several synths. For example, a synth routes its output signal on the bus 4 while another synth can take from there. That is, a hydraulic machinery is connected to a pipe and is feeding the liquid: further along the pipe, a second machine can be joined to the same pipe, and draw the liquid from the junction.

Observation 14
The connecting pipes disperse liquids that not exited through the outlet pipes of the plant in the waste plant. Liquids do not pollute and their dispersion is not relevant

Once routed to a bus, signals are available for computation. If at the end they are not written on a output bus connected to the sound card, there is simply no audible result. In other words, to route a signal to a bus does not require to know in advance what other synths will do of that signal, and the latter may simply not be used anymore. In this sense, what happens to the signal on the bus is irrelevant. This feature ensures the possibility of communication among synths, but without making it mandatory or planned in advance.

---

[6] In this book we will rely terminologically on this categorization, inside/outside or public/private. This is not an official terminology as in SC there is no difference in design or implementation between the two "types".

In summary, the server includes:

- **SynthDefs**: designs for instruments, available for the construction;
- **UGens**: atomic units for signal processing, to be used in the design of instruments;
- **Synths**: real instruments, to be built and used in real-time;
- **Groups**: groups of synths, manageable as a unit;
- **Buffers**: memory blocks, to contain usable data for synthesis/processing;
- **Buses**: channels to route signals inside/outside the server;

What do we do when we use the server? We send commands that allow in real-time to define, build, control, destroy, connect instruments that communicate with each other and that can use available data on the server. It is time to address these issues with the code under our fingers.

5.3 The server

The first thing to do in order to communicate with the server is booting it. The server is represented in the language by the class Server that encapsulates all its functionalities. It is possible to have multiple server instances that can be controlled independently (for example, think of a network of computers, each of which works on a server, all controllable from a single client). By convention and convenience, an instance of Server, directly accessible by the interpreter, is assigned to the environment variable s[7]. So it is not usually appropriate to use this variable s for anything else in SC. The following example shows a set of messages that can be sent to the server to check its operation. Lines 14-15 illustrate the assignment of the default server to another variable (~myServer).

---

[7] There are two types of servers: a "local" and an "internal" one. The latter shares the same memory space with the interpreter and work necessarily on the same machine of the interpreter. In the writer's opinion, this is in contrast with the underlying assumption of the client/server architecture. The internal server was useful in some situations in previous versions of SC: now it is entirely vestigial, and will not be discussed here.

In the IDE, the bottom left window displays some information to monitor the server.

```
1  s // SC replies with "localhost"
2  Server.default ; // the same, assigned to s

4  // Minimal control
5  s.boot ; // boot the server
6  s.quit ; // stop it
7  s.reboot ; // indeed, reboot the server

9  // Gathering information on the generated signals
10 s.scope ; // to visualize signals in time domain
11 s.freqscope ; // to visualize signals in frequency domain

13 Server.killAll ; // in case of zombie processes

15 ~myServer = Server.default ; // assigning the server to another variabile
16 ~myServer.boot ; // same as before
```

Once the server has booted, indications similar to the following are printed on the post window.

- booting 57110: the server is booting and it will receive messages on the port 57110 (an arbitrary, default port number));
- localhost: the default server is in fact a local server;
- lines 4-7: here scsynth interfaces with the operating system and ask for a list of available audio devices, which may include different configurations depending on the operating system, the presence of external devices, etc.;
- lines 9-15: at this point a device is selected, to be actually in use, in the example the built-in sound card. It features two input buses (the microphone here is stereo) and two output ones (the two channels related to the speaker in a standard stereo setup);
- line 17: specifies the sampling rate;
- line 18: the server is ready;
- lines 19-20: a notification has been received back from the server, so there is OSC communication over the network. The last line is not relevant here but everything indicates that it was successful.

```
 1  /*
 2  booting 57110
 3  localhost
 4  Found 0 LADSPA plugins
 5  Number of Devices: 2
 6     0 : "Built-in Input"
 7     1 : "Built-in Output"
 8
 9  "Built-in Input" Input Device
10     Streams: 1
11        0  channels 2
12
13  "Built-in Output" Output Device
14     Streams: 1
15        0  channels 2
16
17  SC_AudioDriver: sample rate = 44100.000000, driver's block size = 512
18  SuperCollider 3 server ready.
19  Receiving notification messages from server localhost
20  Shared memory server interface initialized
21  */
```

Once the server has booted, we can now proceed to exploit its potentialities. Compared to the metaphorical framework introduced above, it is like powering the chemical plant, that is at the moment completely empty. It is up to the user to build and operate the machinery. A useful learning method (useful also in troubleshooting) makes use of s.dumpOSC: with dumpOSC(1) all OSC messages sent to the server are printed on the post window. In this way, the –otherwise hidden– flow of OSC communications from client to server becomes evident. Printing is disabled with dumpOSC(0).

5.4 SynthDefs

First, we need to build the instrument that generates the signal. An example of minimal SynthDef is the following:

```
1  SynthDef.new("bee",
2      { Out.ar(0, SinOsc.ar(567))}
3  ).add ;
```

- SynthDef: is the class on the language side that represents a SynthDef object on the server;
- .new(...): new is the constructor method, which actually builds the SynthDef (returns the SynthDef object). The method new provides a number of arguments. Here, two are specified, for others we leave the default values;
- "bee": the first argument is a string that represents the name of the SynthDef: that is, the name that will be associated with the UGen graph. Synths generated from this SynthDef will be of "bee" type. Here "bee" is a string, but it could also be a symbol (\bee). Note that the SynthDef is not assigned to a variable. In fact, the only utility of a SynthDef is to be sent to the server, not to be manipulated on the language side;
- {Out.ar(0, SinOsc.ar}: the UGen graph is enclosed in braces. All that is in braces in SC is a function, as we already know. Therefore, the UGen graph is described linguistically by means of a function. This UGen graph consists of two UGens, Out and SinOsc: this fact is made explicit by the message *ar that the two UGen receive. In general what answers the *ar or *kr is a UGen. And, in order to generate a real-time signal, a UGen has to receive an *ar or *kr message.

Why a function to describe linguistically the relations between UGens? Remember that a function is an object that returns a value every time it is asked. A synth is then described by a function because at each sample period it is asked to return a value of amplitude, the value of the audio sample. In other words, at every sample period (with the default sampling rate, $44, 100$ times per second), it calculates the value of the function described by the UGen graph.

Figure **5.5** shows, according to the previous conventions, a synth built from the SynthDef "bee". As it may be seen, the resulting signal is "connected" (routed) to the bus 0. In a more usual form, a UGen graph can be described by means of a flow chart, like the graph in Figure **5.6**. An interesting aspect of this graph is that it was generated automatically from the

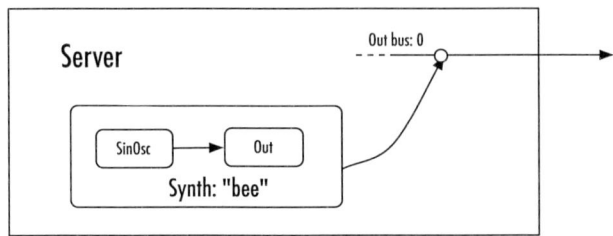

Figure 5.5   Schema of a minimal SynthDef.

SynthDef: therefore it is not a scheme but a real information visualization[8]. Out is the UGen in charge of sending the generated signal to the sound card: without Out there would not be communication with the sound card, and no audible sound. Out has two arguments. The first is the index of the bus on which to send the output signal, the second is the signal to be sent. Thus, Out receives a signal and sends it to the bus audio 0, the first available output channel of the sound card (it is an external, public bus). The signal to be sent comes from SinOsc: it is a case of patching, as the UGen SinOsc is plugged into another Out (see later). SinOsc generates a sinusoidal signal: if frequency and phase are not specified, these will get their default values, respectively 440 (Hz) and 0.0 (radians).

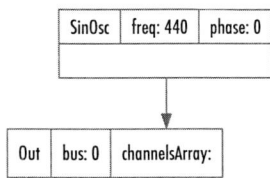

Figure 5.6   Visualization of the UGen graph.

- add: the SynthDef, in itself, is useless if it is not loaded on the server. The method add, defined for SynthDef, tells a SynthDef to "dispatch itself" to the server s. Thus, with add the SynthDef is compiled on the fly in a specific binary format and sent to the server. Remember: everything in a SynthDef

---

[8] The figure was obtained using a modified version of the extensions ("quarks" in SC) dot by Rohann Drape.

consists of specific instructions for the synthesis of the signal, nothing else. By invoking the method add, the SynthDef becomes resident (i.e. available) on the server until this is on.

   In addition to add, many methods defined for the class SynthDef are available. As an example, some allow to write the definition onto a file (this is what the method writeDefFile does): once stored, the SynthDefs will be loaded on the server each time it boots, and therefore will be immediately available for usage. Although at first glance it may seem logical to write libraries of SynthDefs onto files so that they are loaded when the server starts, it is actually a common practice to load them each time. The method add is therefore by far the usual method when working with SynthDef.

   Back to the example, the server now has prepared the project "bee" so that synths of type "bee" may be created. What has been done? A SynthDef was defined (by associating a name and a UGen graph), compiled into a binary format that the server accepts, finally sent to the server that keeps it available in memory until it is running. By turning on the print of OSC messages onto the post window (dumpOSC (1)), once sending the SynthDef we get:

```
1  [ "/d_recv", DATA[106], 0 ]
```

   Once built, the synth "bee" can be thought of as a kind of hardware module without external controls, a black box that can simply be switched on or off. It is worth reviewing the considerations on the functions expressed in the discussion about syntax: a function can communicate with the outside through its arguments. The following example shows a slightly more complex case:

```
1  SynthDef.new(\pulseSine , { arg out = 0, amp = 0.25, kfreq = 5 ;
2      Out.ar(
3          bus:out,
4          channelsArray: SinOsc.ar(
5              freq: kfreq*50,
6              mul: LFPulse.kr(
7                  freq: kfreq,
8                  width: 0.25
9              )
10         )
11         *amp);
12 }).add;
```

Note that in this case the UGen graph associated with the name \pulseSine (this time expressed as a symbol) provides some input arguments, that allow a real-time control. They are out, amp, kfreq. In this way, each synth of type "pulseSine" will offer three controls in relation to the function arguments. Thinking in terms of a hardware module, this time the synth, once built, will present three controls that can be operated from outside (say: three knobs). It is good practice to include whenever possible default values for the arguments of a SynthDef, so that a synth can be created with minimal effort and with "meaningful" parameters. Line 12 closes the UGen graph and sends the SynthDef to the server.

5.5 UGens and UGen graphs

UGens are atomic signal processing units. UGens may have more inputs, but always only one output. A UGen can receive as its input the signal from another UGen: this process is called *patching*. A set of interconnected UGens forms a UGen graph, a graph of UGens, that is, a structure that describes the relations between UGens. Since UGens generate signals, the UGen graph describes the flow of signals that are progressively collected into the output signal. Another metaphor: the UGen graph is the geographical map of a river that collects water from several tributaries, finally ending in the sea.

In the definition of the SynthDef "pulseSine", the program uses formatting and an unusually verbose writing style to make as clear as possible the organization of patching. In the example, lines 2-11 describe the graph of UGens, including an example of patching between Out, SinOsc, LFPulse. Notice that the latter (a square wave generator) updates its values recalculating the output value at control rate, in accordance with the message kr sent to LFPulse.

It is appropriate now to dwell more on a UGen, in particular on SinOsc. To know how the UGen behaves one can obviously read the relative help file. Another option, useful while studying, is to access the UGen definition in the source code.

```
1  SinOsc : UGen {
2      *ar {
3          arg freq=440.0, phase=0.0, mul=1.0, add=0.0;
4          ^this.multiNew('audio', freq, phase).madd(mul, add)
5      }
6      *kr {
7          arg freq=440.0, phase=0.0, mul=1.0, add=0.0;
8          ^this.multiNew('control', freq, phase).madd(mul, add)
9      }
10 }
```

Note that SinOsc inherits from UGen, the generic superclass of all UGens. In addition, it defines only two methods for the class, ar and kr. Leaving aside the last line of each method, we see that the methods provide a number of arguments which can be passed a value. Still, all the arguments typically are provided with a default value. The following three lines are equivalent[9]:

```
1  SinOsc.ar;
2  SinOsc.ar(440.0, 0.0, 1.0, 0.0);
3  SinOsc.ar(freq: 440.0, phase: 0.0, mul: 1.0, add: 0.0);
```

---

[9] Of course it is useless to evaluate them in the interpreter, because they describe objects that are meaningful only if placed in a SynthDef and sent to the server.

The first, in the absence of any indication, uses the default values for the method ar. The second specifies a value for every argument. The third one makes use of keywords, which, as we already know, allow a free order of arguments. The last two arguments are mul and add, and are shared by most UGens: mul is a multiplier for the signal while add is an amount (positive or negative) that is added to the signal. The signal generated by a UGen is by default normalized, its amplitude oscillating in the range $[-1, 1]$ (sometimes, in case of UGens dedicated to unipolar signals, the range is $[0, 1]$). The argument mul defines a multiplier that scales the normalized amplitude, while add is an increase that is added to the same signal. The default values for mul and add are respectively 1 and 0: thus, by default the signal is multiplied by 1 and a value of 0 is added. In this way, it is left unchanged with respect to the default range. For the avoidance of doubt, "multiply" and "add" means that each sample of the signal is multiplied and added to the values specified in the two arguments. Instead in this example:

```
1  SinOsc.ar(220, mul:0.5, add:0) ;
2  SinOsc.ar(220, mul:0.5, add:0.5) .;
```

at line 1 the signal generated by the method ar is multiplied by 0.5 (with 0 added, the latter operation is thus irrelevant): the signal's amplitude will be included in the range $[-1.0, 1.0] \times 0.5 = [-0.5, 0.5]$; at line 2, 0.5 is added to the signal: its amplitude will be included in the range $[-1.0, 1.0] \times 0.5 + 0.5 = [-0.5, 0.5] + 0.5 = [0.0, 1.0]$. It will be unipolar, asymmetrical with respect to 0. To assign mul a constant value $= 0.5$ indicates that each new sample is multiplied by 0.5. In order to generalize the approach, we might think then that mul is a signal, but a constant one. In this regard, the UGen Line provides an example. In the words of the help file, it is a generator of "lines": a line here is a signal that results from "generat[ing] a line from the start value to the end value". The first three arguments of Line are start, end, dur: Line generates a sequence of values linearly from start to end in dur seconds.

In the following code

```
1  SinOsc.ar(220)*Line.kr(0.5,0.5, 10) ;
```

Line generates for a duration of 10 seconds a sequence of values $= 0.5$ (i.e. a progression from 0.5 to 0.5). The output signal from the oscillator SinOsc is multiplied by the output of Line. At each sample period, the sample calculated from the first UGen is multiplied by the sample calculated by the second (which has a constant value of 0.5). Note that the resulting signal is the same as the previous one. The example is meant to demonstrate how a constant (a value) can be thought as a signal (a sequence of values): indeed, in general the interesting feature of Line lies in that the UGen generates values that vary according to a linear progression between the two extremes.

The following example produces a crescendo from silence, because start $= 0$:

```
1  SinOsc.ar(220)*Line.ar(0.0,1.0, 10) ;
```

If, therefore, a constant can be thought of as a signal, then we can think of each value of an argument in a UGen as a constant signal. And therefore, we can consider the case where, rather than a UGen generating constants values, there is a UGen outputting (as usual) a sequence of different values. Patching is precisely the connection of a UGen to an argument of another UGen. The calculation of the signal will thus include the contribution offered by more UGens, organised in a specific configuration of inputs and outputs. The example allows us to understand how the arguments can receive not only constant values but also variable ones, that is, other signals. In other words, signals can change any controllable aspect of other signal generators. Remember that the sampling rate determines the frequency at which the graph is to be traversed. In other words, at every sample period $T = \frac{1}{sr}$ all the UGens recalculate their values, to be used where required.

Figure 5.7 represents the flow chart of the SynthDef "pulseSineGraph". Gray elements describe audio rate flow, the blue ones the control rate flow, the connector boxes represents numeric values. LFPulse works at control rate, while kfreq, amp, out are set at event rate (that is, each time a user changes the parameters). The blocks labelled "* a b" indicate multiplication. In UGens, unspecified values for arguments are indicated by default values.

The multiplier signal is generated by LFPulse, a square wave generator, with a frequency kfreq (thus associated with the frequency of the sinusoid). The output signal from LFPulse is unipolar, that is, in the range $[0, 1]$. If we use

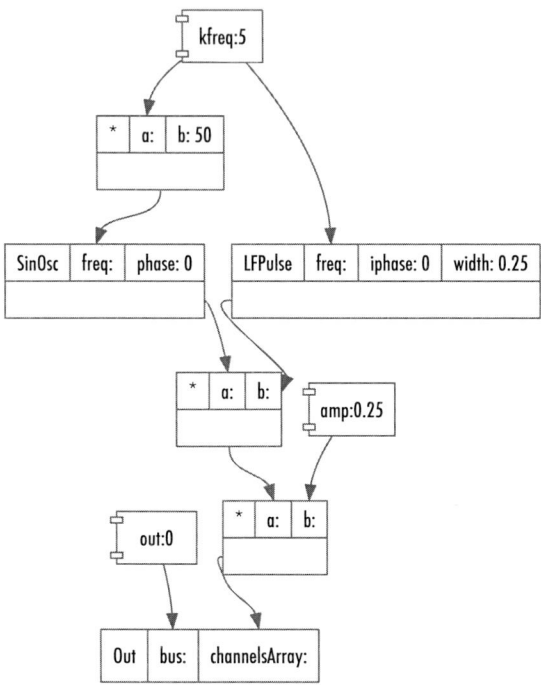

Figure 5.7 Representations of the UGen graph.

a similar signal as a multiplier for another signal, then, when the amplitude is 0.0, the resulting signal has an amplitude 0.0 (silence), when the amplitude is 0.5, the output is the starting signal, scaled by 0.5. In essence, an intermittent opening/closing of the patched signal result. Note that in the example the frequency of the sine wave is related to the intermittent frequency (the more acute the frequency is, the more frequently it is opened/closed).

To summarize: UGens are atomic processing units, described on the language side as classes that receive the (class) methods *ar and *kr, the latter receiving values for the arguments that act as parameters for the UGen. During the real-time synthesis process, UGens generate output signals, that is, sequences of values at audio rate. The values of these arguments may be other signals (patching).

In the example of the SynthDef pulseSine, the writing style was rather verbose, particularly in the extensive use of keywords for UGens. The following two examples are totally identical to the previous version. The first is a form of compact writing that is typical in SynthDefs but that is usually quite complex to decipher for the novice (it requires some exercise). The second version is instead quite lengthy but very clear. Remember, the UGen graph is described linguistically as a function, then, all the considerations we did on functions apply. In the example, two variables are at use (pulser, sine) that are associated with the two signals, with the aim of clarifying the information flow.

```
1  // compact
2  SynthDef.new(\pulseSine , { arg out = 0, amp = 0.25, kfreq = 5 ;
3      Out.ar(out, SinOsc.ar(
4            kfreq*50, mul: LFPulse.kr(kfreq, width: 0.25)
5          )*amp);
6  }).add;

8  // expanded, more verbose
9  SynthDef.new(\pulseSine , { arg out = 0, amp = 0.25, kfreq = 5 ;
10     var pulser, sine;
11     pulser = LFPulse.kr(freq: kfreq, width: 0.25) ;
12     sine =  SinOsc.ar(freq: kfreq*50, mul: pulser) ;
13     sine = sine*amp;
14     Out.ar(bus:out, channelsArray: sine);
15 }).add;
```

As we have seen, UGens are atomic signal processing units. Their number is very large, and it can be checked by evaluating UGen.subclasses.size: in the installation of the writer, the number returned by the evaluation is 305. A rough classification can distinguish:

- **Generation**: signal synthesis, both in deterministic and random fashion (oscillators, noise generators);
- **Processing**: transformation of an input signal (filtering, delaying, reverberation) ;
- **Spatialization**: delivery of an input signal on several channels;
- **Analysis**: signal analysis and feature retrieval;
- **Conversion**: conversion of signals from audio to control and vice versa;
- **Buffer**: buffer allocation and access;
- **Envelopes**: generation and manipulation of envelopes;
- **Triggers**: trigger signals, of various types;

- **Input/Output**: I/O access to the sound card;
- **Info**: information retrieval on audio on the server side;
- **User interaction**: server-side user-interaction (e.g. via mouse);

SC provides the following internal classification for UGens, with their actual amount on the installation taken into account:

Algebraic (5)	GranularSynthesis (32)
Analysis (74)	InOut (20)
Analysis:Synthesis (11)	Info (15)
Base (4)	InfoUGens (1)
Buffer (42)	Maths (17)
Conversion (5)	Multichannel (126)
Convolution (6)	PhysicalModels (2)
Delays (34)	Random (19)
Demand (28)	Reverbs (3)
Deprecated (1)	Synth control (13)
Dynamics (4)	Triggers (31)
Envelopes (8)	Unclassified (4)
FFT (85)	Undocumented (218)
Filters (107)	User interaction (4)
Generators (156)	

5.6 Synths and Groups

Once written and sent to the server the SynthDef, the server simply stores it as a project available to build on the fly a synth, an instrument to produce sound in real-time. The synth, in turn, can be controlled interactively by the user. A typical, minimal program in SC is as follows:

```
 1  // 1. booting the server
 2  s.boot ;
 3  // tracing OSC messagges
 4  s.dumpOSC;

 6  // 2. sending the synthDef
 7  (
 8  SynthDef.new(\pulseSine , { arg out = 0, amp = 0.25, kfreq = 5 ;
 9      Out.ar(
10          bus:out,
11          channelsArray: SinOsc.ar(
12              freq: kfreq*50,
13              mul: LFPulse.kr(
14                  freq: kfreq,
15                  width: 0.25
16              )
17          )
18          *amp);
19  }).add;
20  )

22  // 3. creating a synth
23  x = Synth(\pulseSine ) ;
```

The program is made up of three blocks. First (8-19), the server is booted (and we wait until this to have replied to the client). Then, the block containing the SynthDef has to be evaluated. Again, we wait until the server has replied. Finally, the synth is created (23). The three blocks are, therefore, three moments of asynchronous interaction between client and server. In other words, in principle the client does not know how much time the server will need to respond to its requests. User experience working with SC will show that these waiting times are reduced to some milliseconds, but this does not change the point: interaction between server and client is asynchronous. In case of evaluating all the code in once, the interpreter would execute all the sequence of expressions as fast as possible. But, then, the server would still be booting when the SynthDef is sent, and thus the latter would not be received. Once such an operation has failed, the requesting for a synth would lead to:

```
1  *** ERROR: SynthDef pulseSine not found
2  FAILURE IN SERVER /s_new SynthDef not found
```

An error message that indicates precisely that the SynthDef request to build a synth is not available on the server.

Going back to the program, as can be imagined, SC provides a class named Synth that encapsulates all commands pertaining to a synth on the server. The constructor method new, as usual omissible and here omitted, includes as an argument a string indicating the SynthDef from which to manufcture the synth ("pulseSine"). Once created, the synth is immediately activated (= it plays). This behavior is useful because a synth can be thought of as an instrument (a synthesizer) but also as a sound event (e.g. a "note"): considering this second option, it becomes obvious that creating a synth is like generating a sound event. Once the synth has been created, the GUI text view dedicated to the server in the IDE is updated, and the values of the fields "u" (active UGens) and "s" (active synths) vary accordingly.

Before going into the details of synth control, a crucial question has to be answered: how can we stop the sound? The "life-saving" command in the IDE is Language→Stop, that interrupts every generation process (both sound synthesis and scheduling)[10].

Turning again to the previous program, but this time activating the printing of OSC messages with dumpOsc, we get the following post window:

```
1  a SynthDef
2  [ "/d_recv", DATA[343], 0 ]
3  Synth('pulseSine' : 1000)
4  [ 9, "pulseSine", 1000, 0, 1, 0 ]
5  [ "/g_freeAll", 0 ]
6  [ "/clearSched", ]
7  [ "/g_new", 1, 0, 0 ]
```

---

[10] The keyboard shortcut depends on the platform, typically "CTRL or APPLE + .".

Without going into too much detail, lines 1 and 2 indicate the receiving of the SynthDef (in the form of a standard answer by the interpreter, including the OSC message), then (3-4) the creation of the synth, finally (5-7) the sequence of OSC messages that result from stopping the sound. As for the synth, an observation can be added: the number 1000 is an identifier that the server progressively assigns to synths, starting from 1000 (the previous indices are reserved for internal use). If another synth is created (without calling Stop), it will be identified by the index 1001. In any case, while coding synths are usually referred not directly by their ID, rather by assigning an instance of the class Synth to a certain variable. This is an example of the usefulness of abstracting server constructs into language classes: by encapsulating all the functionalities of a synth into an object (including its ID), the Synth object manages all the required bookkeeping offering the user a unified interface. Lines 5-7 indicate that as a result of the Stop command, the server was required to remove all the synths. Here, g refers to a group that is active by default (see later).

The next example, abundantly commented, assumes that the SynthDef pulseSine is still available on the server. Line 4 shows how to instantiate a synth by immediately setting the values of some of the arguments for the UGen graph function (the arguments would otherwise receive their default values). The syntax requires to provide an array alternating the names of the arguments and their relative values. Lines 5-6 show how to play/pause a synth by means of the method run. Lines 8-9 show how to change the values of some arguments of the UGen graph function by using the method set that can receive a list of arguments/values to set the desired parameters. Line 11 shows how to eliminate (technically, "deallocate") a synth through free. Finally, lines 14-15 show how to create a synth without having it running immediately by using the constructor method newPaused: the synth can be activated at a later time (line 15).

```
1  // Arguments and synth control

3  // intreactive control, you need to evaluate each line (= performance)
4  x = Synth(\pulseSine , [\kfreq , 14, \amp , 0.7]) ;
5  x.run(false) ; // = press the pause button on the synth
6  x.run(true) ; // = x.run, true is the default value

8  x.set(\kfreq , 15, \amp , 0.125) ; // control of arguments
9  x.set(\kfreq , 20) ; // control of arguments, just one

11 x.free ; // deallocation: synth is eliminated

13 // from beginning but starting with a paused synth
14 x = Synth.newPaused(\pulseSine , [\kfreq , 5, \amp , 0.5]) ;
15 x.run ;
```

The code above is an interactive session with the interpreter that controls in real-time the server to process audio. Even if minimal, it is an example of "live coding", that is, of live, interactive programming. In fact, in SC it could be said that the user (indeed at different degrees) always works in a live coding fashion.

Synths can be coordinated in a group: a Group is simply a list of synths to which is possible to send the same message. Groups and synths have substantially the same interface, i.e. the basic methods of synths (which we have already discussed) also apply to groups. Consider the following example:

```
1  SynthDef(\sine , {arg  freq = 100; Out.ar(0, SinOsc.ar(freq))}).add ;
2  SynthDef(\pulse , {arg freq = 100; Out.ar(1, Pulse.ar(freq))}).add ;

4  s.scope ;
5  g = Group.new ;
6  x = Synth(\sine , [\freq , 200], target:g) ;
7  y = Synth(\pulse , [\freq , 1000], target:g) ;

9  g.set(\freq , 400) ;
10 x.set(\freq , 1000) ;
11 g.free ;
```

Lines 1-2 define two minimal SynthDefs, respectively, a sine wave oscillator and a square wave one. Both SynthDefs expose as an interface the argument freq that controls the frequency in both the oscillators. The two oscillators route the signal respectively to the left and right channel through the argument passed to Out that specifies the bus (0 vs 1). Calling s.scope makes more clearly observable what happens (4). A group is then created through the class Group. Two synths are then created (lines 6-7), without anything special except for the specification of the argument target that allows to assign a synth to a group, here the g. Once the group has been created, it becomes possible to control the associated synths in one shot. Line 9 sends to all the synths belonging to the group g a message that sets freq = 400. It should be noted that the syntax is the same as for synths[11]. All the synths in the group that provide a freq argument (both in our case) responds to the message sent to the group. The association to a group does not prevent the control of individual synths (10). Note that the entire group can be deallocated in one message, that is, all the synths that belong to it (11). Groups are useful to impose modularity (audio tasks can be split among various synths) but at the same time coordination: a classic example is scaling the volume of many synths together. Moreover, a group can in turn contain other groups. When a synth is created without specifying the target group, it is aggregated to the so-called "default group".

5.7 A theremin

The well-known theremin is an electronic instrument invented by Russian Léon Theremin in which two antennas control the amplitude and the frequency of an oscillator. The performer plays without actually touching the antennas, as the control is achieved as a function of the proximity of the hands to the same antennas. Consider then the following example (commented to emphasize again some issues about SynthDefs):

---

[11] In fact, both Synth and Group are subclasses of the more general class Node, and by the way Node's help file contains relevant information for both synths and groups.

```
1  SynthDef.new(\theremin ,   // a symbolic name
2      // the ugen graph follows
3      {
4      // we always need the Out ugen
5      Out.ar(
6          0, // bus index
7          // then the signal generated by the digital oscillator
8          SinOsc.ar( // SineOsc is the oscillator UGen
9              freq: MouseX.kr(200, 5000, 1), // argument: frequecy
10             mul: MouseY.kr(0,1))) // argument: a multiplier for amplitude
11 }).add ;
```

The structure is very simple: a sine generator is controlled in its amplitude and frequency by two peculiar UGens, MouseX and MouseY. These are two "pseudo-UGens", as they intercept the mouse (which is on the client side, evidently) to generate control signals. They are very useful to quickly check the behavior of other UGens by exploring interactively a range of values. The two UGens refer to the mouse position on the horizontal and vertical axes of the screen, and they allow to map the x/y coordinates in the range defined by minval and maxval. Also, the type of curve that connects the two extremes can be selected (warp). In the example, the frequency range (handled by MouseX) is $[200, 5000]$ (it refers to Hz), while the amplitude range (handled by MouseY) is $[0, 1]$ (normalized amplitude). The curve is linear (0) for the amplitude and exponential for frequencies (1), in order to better account for the perception of pitches[12]. By using the mouse, the basic operations of a theremin can be simulated.

The UGen graph is shown in Figure **5.8**. The graph allows to enter into some details concerning the arguments mul and add, share by most UGens. Internally, there is actually a specialized UGen, MulAdd, which operates very efficiently. The UGen is not intended to be explicitly invoked by the user, rather it is an internal implementation solution. But the graph, being generated by inspecting the structure of the actual SynthDef, detects it and draws it (the block labelled *).

---

[12] The argument lag, usually left unset, is a delay that prevents an excessive time sensitivity to the mouse jitter.

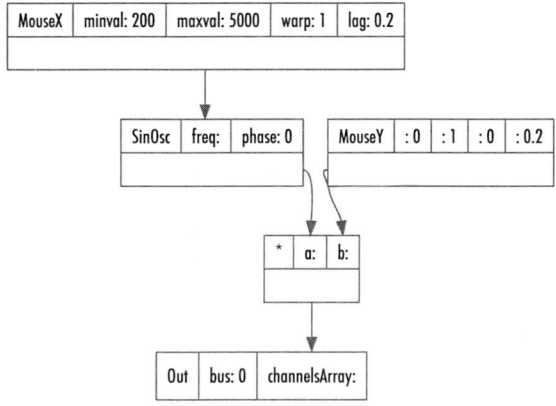

Figure 5.8   UGen graph of the theremin.

To see what happens in terms of waveform and spectrum, it is possible to evaluate s.scope and s.freqscope.

5.8 An example of real-time synthesis and control

It is now worth discussing an example that combines graphical control and audio synthesis: a first, minimal program that relates structurally the client and the server.

```
 1  Server.local.boot ;      // same as s.boot

 3  (
 4  // audio
 5  SynthDef.new("square", { arg out = 0, freq = 400, amp = 0.75, width = 0.5;
 6      Out.ar(out, Pulse.ar(freq, width: width, mul: amp));
 7  }).add;
 8  )

10  (
11  // variables
12  var aSynth, window, knob1, knob2, button;

14  aSynth = Synth.new(\square ); // the synth

16  // GUI: creation
17  window = Window.new("Knob", Rect(300,300,240,100));
18  window.front;

20  knob1 = Knob.new(window, Rect(30, 30, 50, 50));
21  knob1.valueAction_(0.25);

23  knob2 = Knob.new(window, Rect(90, 30, 50, 50));
24  knob2.valueAction_(0.3);

26  button = Button.new(window, Rect(150, 30, 50, 50)) ;
27  button.states = [      // state array
28               [ "stop", Color.black ], ["start", Color.red]] ;

30  // GUI: controlling audio
31  knob1.action_({arg me; aSynth.set(\amp , me.value) });
32  knob2.action_({arg me; aSynth.set(\freq , me.value.linlin(0,1, 200, 2000)) });
33  button.action = ({ arg me;
34      var val = me.value.postln;
35      if (val == 1) { aSynth.run(false) } { aSynth.run }
36  });

38  window.onClose_({aSynth.free})
39  )
```

The SynthDef is a simple wrapper for the UGen Pulse that generates square waves, and permits to control its parameters: width is the duty cycle, i.e. the ratio between positive and negative part of the signal (in the range [0, 1], where the default is 0.5, indicating an equal proportion). The block 10-38 is instead deputed to instantiate the synth and its control via GUI. Line 14 creates a synth, and assigns it to the variable aSynth.

Not much left to say about creating the GUI elements. Two knobs are placed inside a window, according to the technique already discussed. The actions related to the rotation of the knobs and button are more interesting.

Line 31 defines the connection between the GUI controller Knob1 and the audio synthesis. The action is associated with the knob through the method action. Every time the value of the knob changes, the set method is invoked on the synth aSynth, that assigns the specified parameter (amp) a value, here v. value. Since a Knob object generates values in the range [0, 1], i.e. in the same amplitude range of a normalized audio signal, the knob's range can be used directly to control the amplitude (i.e. its "volume"). Similarly, line 32 assigns the argument freq the value of the knob knob2. When frequencies come into play, the knob's default range [0, 1] is no more appropriate. The method linlin easily allows to linearly map an input range (defined by its first two arguments) to an output one (defined by the second two arguments), here [200, 2000]. Similarly, it would have been possible to use the method linexp that has the same syntax of linlin, but provides an exponential mapping. The situation is represented in Figure **5.9**.

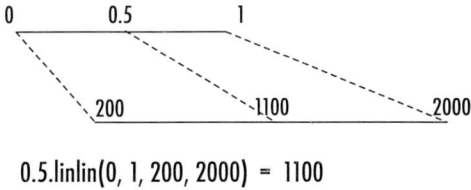

0.5.linlin(0, 1, 200, 2000) = 1100

Figure 5.9 Linear interpolation with linlin.

The GUI button button is assigned an action too, according to a syntax similar to that of Knob: me.value provides access to the state of the button (more precisely, to the index of the state in the array of states button.states). Each time the button is pressed, the action function is evaluated. The action checks the state of the button through the conditional (35). If the value val of the button (printed on the post window) is 1, the method run (false) is called for aSynth, pausing the synth. In the opposite case (thus, val = 0), the synth is activated (aSynth.run). When running the code, the synth is up and the state is 0. After the first button press, the state becomes 1, the GUI is updated and the conditional branch that contains aSynth.run(false) is evaluated.

Lines 21 e 24 set the value of the two knobs by the method valueAction that not only modifies the graphic element but evaluates the associated action by passing a value, in this way keeping in sync GUI and value. Finally, line 38 exploits the method onClose that associates to the window an action to be performed when the former is closed: here, the deallocation of the synth aSynth.

5.9 Expressiveness of the language: algorithms

There are good chancec that the SuperCollider language would not be exactly intuitive for a musician, yet it is very expressive. For example, consider the following program:

```
1  (
2  SynthDef(\sine , {|freq = 440, amp = 1, gain = 1|
3      Out.ar(0, SinOsc.ar(freq, mul:amp*gain))}).add ;
4  )

6  (
7  var base = 20, keys = 88 ;
8  var  synths, group;
9  var window, step = 15, off = 20, len = 80 ;
10 group = Group.new;
11 synths = Array.fill(keys, {|i|
12     Synth(\sine , [\freq , (base+i).midicps, \amp , 0], group)});
13 window = Window.new("sliderPiano", Rect(10, 10, keys*step, off+len+off+10))
14     .front ;
15 keys.do{|i|
16     StaticText(window, Rect(i*step, 0, step, step))
17         .string_((base+i).midinote[0..1]);
18     Slider(window, Rect(i*step, off, step, len))
19         .action_{|me|synths[i].set(\amp , me.value/keys)}
20 };
21 Slider(window, Rect(0,step+len+10, step*keys, off))
22     .action_{|me| group.set(\gain , me.value.linlin(0, 1, 1, 9.dbamp))};
23 window.onClose_{group.free} ;
24 )
```

The code creates a "piano keyboard" of sliders, each connected with a sinusoid generator whose frequency is related to the note indicated at its top. The horizontal slider acts as a sort of "gain" control, increasing the volume of all the generators. Three things are worth considering:

- only 2 lines (11-12) result in the creation of 88 synths: to be precise, it is actually one single expression;
- similarly, lines 15-20 generate, again by means of a single expression, 88 sliders related to the synths;
- the extensive use of variables allows to parametrize the whole code. For example, try base = 20 and keys = 12. You will get the octave of middle C.

The code exploits constructs that have been thoroughly discussed hitherto. The SynthDef is very simple. The only relevant issue is the multiplier amp. It can be thought of as a gain, hence the name gain. Figure **5.10** shows the structure of the SynthDef as it appears to the server.

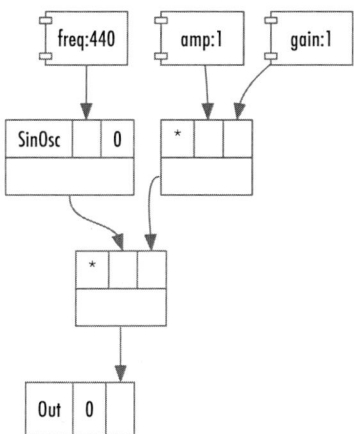

Figure 5.10   Structure of the SynthDef sine.

The variables base and keys respectively indicate the lowest note and the number of notes. Here pitches are expressed following the midi protocol that assigns a progressive integer number every note, arbitrarily assuming 60 for the middle C. Instead the parameters step, off, len are related to graphic parameters, in

particular step is the step upon which all the GUI elements have been built. After defining a group group, the array synths has been created as a container for
the synths, their number being indicated by keys. The function passed to the
method fill allows to exploit the counter i. The synths will have a frequency
with progressive increase of 1 (semitone), starting from basis.

All the synths starts with amplitude = 0. They are all active, as they are
generating signals, but with zero amplitudes. The synths are aggregated in the
group group (12). The width of the main window is calculated according to
the number of synths (multiplied by step) (13). At this point, a number keys
of graphic elements are created in pairs (Sliders and StaticTexts, i.e. labels).
Note that the position of the objects is referred to the counter i and to step.
The class StaticText let the user define a static label element, with a string set
by the method string. Here, the text to be written is the note name, which is
retrieved through midinote, and by taking only the first two characters ([0..1])
of the returned string, for sake of compactness (17). The action associated to
the slider (19) first retrieves a synth from synths at the relative index i, then
assigns it the amplitude obtained from the value of the slider (19). For reasons
of safety, the amplitude is divided by the number of notes, to avoid any possible
clipping (according to a method that we have already described). The slider
(21-22) exploits the group organization, and sends to all the synths of the group
a value that is used as a multiplier for amp (19). Here the gain is expressed in
decibels (9.dbamp) and converted into amplitude. Note that we need to collect
the synths into an array (synths) exactly because they need to be addressed
in relation to sliders. On the contrary, the following GUI elements are simply
printed on the screen and no longer modified, so it is useless to collect them
into an array for future reference. The same applies to the slider that controls
the gain. A final control element is associated with the window. When it closes,
the whole group is deallocated (23). Note the compactness and the flexibility
of the code, that produces a real GUI-controllable, oscillator bank.

5.10 Expressiveness of the language: abbreviations

When there is sound from the server there are necessarily two elements:

- a SynthDef that specifies an organization of UGens;

- a synth built from it.

Law of excluded middle. Yet, especially while exploring the help files for UGens, for example SinOsc, expressions of this kind pop up very often:

```
1 { SinOsc.ar(200, 0, 0.5) }.play;
```

If the expression is evaluated, a sine wave starts. Yet there is no SynthDef, no synth, there is only one function (which, moreover, does not contain the UGen Out) on which the method play is called. Indeed, what happens during the interpretation is a complex process. As a result of the message play, the interpreter:

- creates a SynthDef;
- defines an arbitrary name and associates it with the UGen graph func defined by the function;
- add to the graph a block of type Out.ar(0, func) which routes the signal to the bus 0;
- associates the name to the graph and sends the SynthDef to the server;
- waits until the SynthDef is compiled;
- creates a synth from it.

By evaluating the code, on the post window we get something like this:

```
1 Synth('temp__0' : 1000)
```

It can be seen that the SynthDef has a name 'temp__0'. Graphically inspecting the graph of the SynthDef results in Figure **5.11**. The only new fact is that here Out (stroked in yellow) is not audio- or control rate, but instead gets a ir rate (i.e. "initial rate"). This indicates that the UGen cannot be changed anymore (it is static), a solution computationally more efficient as the user has not requested control over it (s/he has not included Out at all).

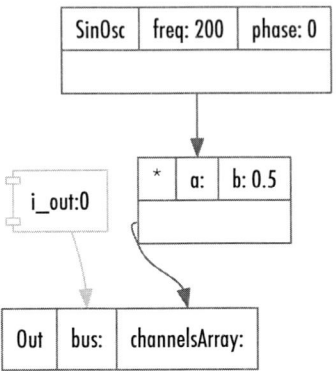

Figure 5.11   Graph of the temporary SynthDef.

The construct {...}.play is very useful for at least two reasons. First, it enable us to quickly experiment with UGens, skipping during the prototyping stage the more structured but cumbersome path of writing a SynthDef and allocating a synth. Secondly, it makes it easy to create a synth. In fact, the evaluation of such a function returns a reference to the created synth. In the following example:

```
1  x = {|fr = 1000| SinOsc.ar(fr, 0, 0.5) }.play ;
2  x.set(\fr , 400) ;
3  x.free ;
```

The environment variable x is associated with the synth returned by the function. From now on, the synth can be controlled as usual. Figure **5.12** visualizes the relative UGen graph, that also shows the accessible argument.

Note that also the output bus can still be controlled. In the following example (which introduces the UGen Saw, specialized for the generation of sawtooth waves), the function argument out specifies the value for the argument out in the UGen Out. Being explicitly introduced in the function, it can be controlled in the resulting the synth.

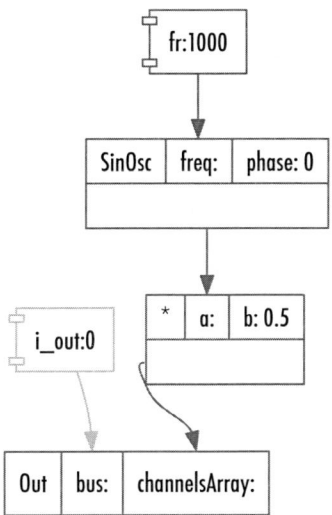

Figure 5.12   Graph of the temporary SynthDef, with argument.

```
1 x = {|fr = 1000, out = 0| Out.ar(out, Saw.ar(fr, 0.5)) }.play ;
2 x.set(\out , 1) ; // now on right chan, bus 1
3 x.free ;
```

The graph is shown in Figure **5.13**, where it can be seen that i_out is created anyway, but not used.

The method play has been deliberately introduced very lately, as it implies a lot of behind-the-scenes activity. Though very useful, it might actually be a confusing shortpath for the SuperCollider novice. It will be very used onwards, nevertheless.

5.11 Conclusions

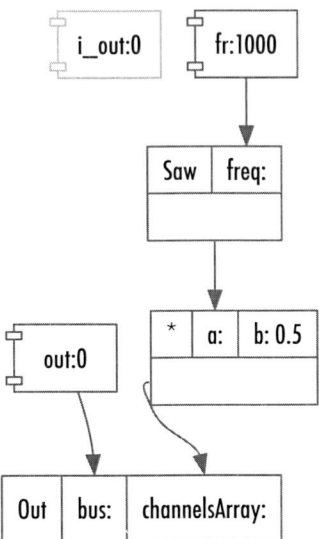

Figure 5.13   Graph of the temporary SynthDef, with Out.

Once introduced, the structure of the server (if one considers Figure **5.4**) is not all that complex. There are some elements (six to be exact) and a communication protocol to control them by sending messages. What should be now very clear is the following statement: *if there is sound in real-time in SuperCollider, then there is a synth on the server, that has been manufactured from a SynthDef*. To clearly keep in mind what is on the server and what is on the client is a basic requirement to operate effectively with SuperCollider.

6 Control

The discussion on the ontology of the server is not yet complete: we still have to introduce Bus and Buffer. Before discussing them, however, it is worth revisiting Envelopes. They are server-side resources that facilitate synthesis and signal processing and, therefore, can be broadly understood as relevant to a chapter about control.

6.1 Envelopes

As already demonstrated, we can forge dynamic control of a signal using Env. But an instance of an Env is not really a signal, but rather an abstraction consisting of times, values and descriptions that can be expanded (or condensed) when desired/needed. While Env objects can be converted to signals by means of the asSignal message, they are typically wrapped with a specialized envelope-generator UGen, namely EnvGen, which generates the appropriate signal within a synthesis graph. In that vein, Env's times argument should not be understood as absolute time, but instead should be thought of in respect to the wrapper EnvGen's state. Consider the arguments of EnvGen's *ar method (which are the same as those of *kr):

```
1  *ar { arg envelope, gate = 1.0, levelScale = 1.0, levelBias = 0.0,
2      timeScale = 1.0, doneAction = 0;
```

As explained in the help file, envelope is an instance of Env; levelScale, levelBias, and timescale are transformation operators that operate on the values represented in an instance of Env. Consider the following example:

```
1  e = Env.new(
2      levels:[0.0, 1.0, 0.5, 0.5, 0.0],
3      times: [0.05, 0.1, 0.5, 0.35]
4  ).plot ;
```

- levelScale: all elements in the envelope's array of values are multiplied with this number. The default value is 1.0, thus leaving the original values unchanged.
- levelBias: this number is added to the envelope's values. E.g., if the latter are into the $[0.0, 1.0]$ range, they will be transposed to $[0 + levelBias, 1.0 + levelBias]$. The default value is 0.0, again leaving the original values unchanged.
- timescale: all elements in the envelope's array of durations are multiplied by this number. The overall duration of the generated envelope is, therefore, $timeScale \times times$. As far as the above example is concerned, the sum of all durations equals 1; the duration of the generated envelope is, then, $1 \times timeScale$. The default value is 1.0, so that the original values are preserved if no value is specified.

Therefore it is possible to stretch/shrink an envelope in both the time and level dimensions. Accordingly, it is sometimes convenient to define our envelopes within nominal ranges —values should be within a normalized $[0.0, 1.0]$ range and the sum of all durations should be exactly 1 — and then rely on the above arguments to transpose them as needed/desired. This is demonstrated in the following piece of code (which is based on an example from the EnvGen's help file):

```
1 | // explicit multiplication of signals
2 | { EnvGen.kr(Env.perc, 1.0, doneAction: 0)
3 |      * SinOsc.ar(mul: 0.5) }.play ;

5 | // effects of timeScale
6 | { EnvGen.kr(Env.perc,  1.0, timeScale: 10,   doneAction: 0)
7 |      * SinOsc.ar(mul: 0.5) }.play ;

9 | // multiplication using mul
10| { SinOsc.ar(mul: 0.1
11|      * EnvGen.kr(Env.perc,  1.0, timeScale: 10,   doneAction: 0) ) }.play ;
```

The effect of the timeScale argument is evident when comparing the first two examples—the original duration is scaled by a factor of 10 in the second. This serves, as well, as a typical example for implementing an amplitude envelope: by means of multiplying the original audio signal with the one generated by EnvGen. Amplitude envelopes are unipolar signals, so when they are multiplied with bipolar audio signals, the result will be an amplitude modulated versions of the latter. An alternative way to arrive at the same result is the one illustrated in the third example, where the envelope signal is used as a value for SinOsc's mul argument. Try replacing play with plot in the above examples to visualize the resulting signals.

The remaining two arguments, namely gate and doneAction, should be understood with respect to a rather complicated problem: how to apply a finite signal (the envelope) to an infinite one (e.g. the one resulting from a SinOsc).

- gate: the envelope signal is generated when a *trigger* occurs in the gate argument. Triggers are to be understood as signals that "cause something to happen". They behave like photocells which emit a signal whenever they detect light. In SuperCollider, triggers are simply some transitions from $<= 0$ to a > 0 value. An EnvGen will generate a signal of constant zeroes (or of any other default value) until a trigger is registered; it is only then that the actual envelope signal is generated. In all the above examples the gate value has been set to 1.0 (the default value), so that the envelopes are immediately generated. Consider, however, the first example in the following block of code: here gate is 0 and the envelope is, in fact, never triggered. If 0 is replaced with some signal, the envelope will be activated when a trigger happens, and in fact, reactivated every time signal in the gate argument transitions

from a $<=0$ value to any positive one —note that transitions from $>=0$
to 0 do not work as triggers. Consider the second example in the following
block of code, where a sinusoidal signal of frequency 4 modulates the gate.
In a Cartesian representation of the signal, each time there is a cross to the
positive plane in the y-axis (that is, whenever the output of the function is
greater than 0), a trigger is registered. In this case, this occurs 4 times per
second. In the last example, a MouseX is used to generate triggers. The horizontal axis of the screen is mapped to a $[-1, 1]$ range, 0 being, therefore, in
the middle of the screen. Whenever the mouse cursor crosses the middle
of the screen from left to right (that is, whenever there is a transition from
0 to a positive number), the envelope is triggered. All in all, gate is Super-
Collider's way of handling the activation of envelopes —much in the same
fashion that commercial synthesizers adopt the solution of key-triggering.

```
1  // gate = 0
2  { EnvGen.kr(Env.perc, 0.0, doneAction: 0)
3      * SinOsc.ar(mul: 0.5) }.play ;

5  // controlling gate
6  // with a signal
7  { EnvGen.kr(Env.perc, SinOsc.kr(4), doneAction: 0)
8      * SinOsc.ar(mul: 0.5) }.play ;
9  // with mouse
10 { EnvGen.kr(Env.perc, MouseX.kr(-1,1), doneAction: 0)
11     * SinOsc.ar(mul: 0.5) }.play ;
```

- doneAction: consider a percussive-style envelope: what happens when it is
  over? What about the corresponding synth? Well, it simply remains active
  unless explicit de-allocation takes place. That is to say that EnvGen (silently)
  keeps generating the last computed value. Therefore, when the latter is 0,
  such as e.g. in the previous examples, the envelope does end, acoustically
  speaking, but the EnvGen keeps outputting the last value in the envelope and
  the enclosing synth remains active. Consider the first line of the following
  block of code: the envelope's last computed value is 1 and, accordingly, the
  synth will keep producing sound until we manually free it. Even if we can
  not hear it, this is exactly what happens when the envelope ends with a value
  of 0: it is still up to the user to manually free the synth. Note that, even if they

merely produce silence, running synths are a significant resource leak since
they reserve physical memory and they keep the CPU busy processing the
entire synthesis graph. The method doneAction is meant as a way to easily handle automatic de-allocation: with doneAction the user may specify
what should be done when an envelope is over. Currently there are 14 possible choices, such as e.g. doing nothing (doneAction: 0, the default value),
automatically deallocating the enclosing synth when the envelope is over
(doneAction: 2), and others. Done-actions that free the synth can be visually monitored in the Server's GUI, which shows a count of all active synths
at any given time —notice the change in the count when evaluating line 12
in the example below. Of course, automatically freeing a synth means that
we have to create a new one if we want the same sound again (it is no longer
possible to re-trigger the envelope).

```
1  { EnvGen.kr(Env([0,1], [2], doneAction: 0)) * SinOsc.ar(mul: 0.1) }.play ;

3  (
4  SynthDef.new("sinePerc",
5      { Out.ar(0,
6          EnvGen.kr(Env.perc, 1.0, doneAction: 2)
7          *
8          SinOsc.ar(mul: 0.1))
9  }).add ;
10 )

12 Synth.new("sinePerc") ;
```

Consider the following example where the mouse is used to trigger the
synth. With doneAction: 0 (line 1) the synth remains allocated after the envelope is over and, hence, it can be re-triggered. Yet, if doneAction: 2 (line 3),
the synth is automatically freed and is no longer able to interact with the mouse
or, really, in any other way.

```
1  {EnvGen.kr(Env.perc, MouseX.kr(-1,1),doneAction:0) * SinOsc.ar(mul:0.1)}.play;

3  {EnvGen.kr(Env.perc, MouseX.kr(-1,1),doneAction:2) * SinOsc.ar(mul:0.1)}.play;
```

These two different settings for doneAction essentially represent two different conceptualizations of what a synth is. On the one hand, a synthesizer is supposed to be an instrument; so, once created, it should remain ever-available for the user to interact with it—much alike a hardware synthesizer which is not supposed to self-destruct after the action of some button is over. On the other hand, a synth can be thought of as a sound event per se, rather than its generator. In that case it makes perfect sense to automatically deallocate any associated process once a sonic event is over.

In the following example, a minimal instrument with a GUI is created: an one octave long keyboard. The code block includes just two expressions; the first creates the GUI window (line 1), the second is an iterative structure which does all the rest, that is, 12 times (one per note) do:

1. define a note by means of adding the iteration counter to midi note 60 (the middle C);
2. make a button for the note and place it sequentially from left to right in the parent window (line 4).
3. define the button's state: the name of the corresponding note is retrieved by accessing an array containing note names by the index resulting from modulo 12 on the the note number (6, 7). It is then set in white color (line 8) on a varying background —the color hue follows the counter (line 9);
4. associate each button with the creation of a synth —of an envelope-controlled square wave, the frequency of which is that of the desired note. There is no need to assign the synth to any variable; it will automatically free itself when the envelope is over (doneAction: 2). A synth here is to be understood simply as an abstraction of a singleton note.

```
1  w = Window("miniHarp", Rect(100, 100, 12*50, 50)).front ;
2  12.do{|i|
3      var note = 60+i;
4      Button(w, Rect(i*50, 0, 50, 50))
5      .states_([[
6      ["C", "C#", "D", "D#", "E", "F", "F#", "G", "G#", "A", "A#", "B"]
7          [note%12],
8          Color.white,
9          Color.hsv(i/12, 0.7, 0.4)]])
10     .action_{
11         {Pulse.ar(note.midicps)*EnvGen.kr(Env.perc, doneAction:2)}.play
12     }
13 } ;
```

6.2 Generalizing envelopes

Although their most common application is to modulate amplitude, envelopes can be used to control any kind of parameter. Env objects are very similar to the so-called tables —tabulated breakpoints that describe some data distribution. EnvGen is, then, a reading module for such kind of tables, whatever value they may hold. Consider the following piece of code:

```
1  {
2      var levels, times, env ;
3      levels = Array.fill(50, { arg x ; (x%7)+(x%3)}).normalize ;
4      times = Array.fill(49, 1).normalizeSum ;
5      env = Env.new(levels, times) ;
6
7      Pulse.ar(EnvGen.kr(env, 1, 100, 200, 20, 2))
8  }.play
```

There are several things to consider herein. An envelope is defined in lines 2–5, which comprises 50 break-points, representing the amplitude (line 3), interspersed by 49 points, representing in-between durations (line 4). `levels` is an array containing the results of 2 different iterative loops where the modulo (%) operation is applied to an incrementing counter (x). The modulo operator makes it easy to produce complex, albeit not random, curves. The resultant curve is shown in Figure **4.15**.

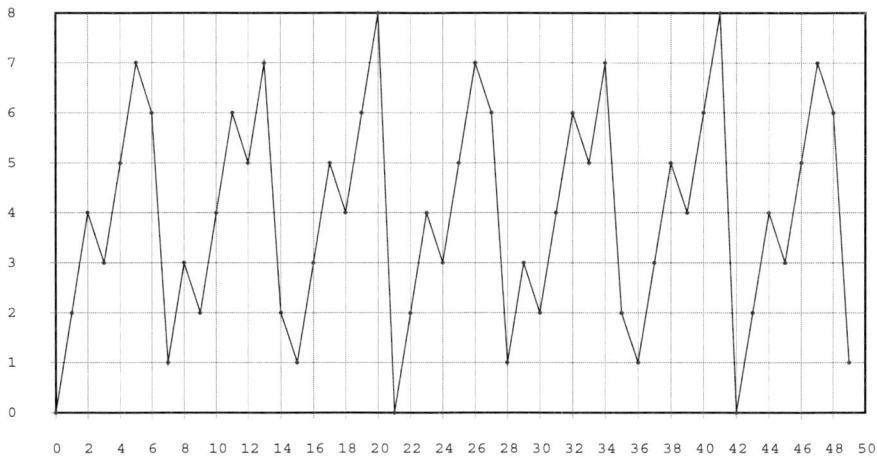

Figure 6.1 Array built by the modulo operator.

The period of each cycle is given by 7 and 3, respectively, so that they are in phase at their minimum common multiple (21). Accordingly, the maximum of their sum is given by $6 + 2 = 8$ (the sum of their local maxima) and the minimum by $0 + 0 = 0$. The `normalize` method then scales all values to a nominal range of $[0.0, 1.0]$. `times` is defined as an array filled with the constant 1, this way suggesting that all amplitude break-points are equally-spaced in time. The `normalizeSum` method returns `array / array.sum`, so that each element of the new array is the original value divided by the sum of all elements in the original array. The effects of `normalize`, `sum`, `normalizeSum` are illustrated in the following code snippet.

```
1  [1,2,3,4,5].sum // sum
2  15

4  [1,2,3,4,5].normalize // max and min in [0, 1]
5  [ 0, 0.25, 0.5, 0.75, 1 ]

7  [1,2,3,4,5].normalizeSum // sum of the elements = 1
8  [ 0.066666666666667, 0.13333333333333, 0.2, 0.26666666666667,
9     0.33333333333333 ]

11 [1,2,3,4,5].normalizeSum.sum // summing the normalized sum = 1
12 1
```

levels and times are, in this way, converted to a normalized form which is easier to deal with. The idea is to use EnvGen's parameters to scale the resulting envelopes and to eventually modulate the frequency of Pulse. Accordingly, levelScale is set to 100 and levelBias to 200, so that the envelope is transposed to the [200, 300] range. timeScale is, then, set to 20 so that its total duration is 20 seconds. Note also that doneAction is set to 2, so that the synth will be automatically de-allocated when the envelope is over. This envelope is a ramp and will, therefore, generate a glissando effect. A portamento can be achieved, instead, by means of wrapping the envelope with a Latch UGen.

```
1  {
2      var levels, times, env ;
3      levels = Array.fill(50, { arg x ; (x%7)+(x%3) }).normalize ;
4      times = Array.fill(49, 1).normalizeSum ;
5      env = Env.new(levels, times) ;

7      Pulse.ar(Latch.kr(EnvGen.kr(env, 1, 100, 200, 20, 2), Impulse.kr(6)))
8  }.play
```

Latch is an implementation of the classic *Sample & Hold* module. On each registered trigger Latch samples its input signal, and keeps generating that value on its output until a new trigger occurs. In the above example, an Impulse —a train of distinct samples— is used to generate triggers, at a frequency of 6. Thus, every $T = \frac{1}{6}$, Impulse generates a unique sample of amplitude mul.

Remember that it does not really matter what value a triggering signal has, as long as there is a transition from $<= 0$ to > 0. Accordingly, the 20-seconds envelope curve is sampled every $T = \frac{1}{6} = 0.16666666666667$ seconds. The resulting effect is further emphasized if levelScale is set to 24 and levelBias to 60:

```
1  {
2      var levels, times, env ;
3      levels = Array.fill(50, { arg x ; (x%7)+(x%3) }).normalize ;
4      times = Array.fill(49, 1).normalizeSum ;
5      env = Env.new(levels, times) ;
7      Pulse.ar(Latch.kr(EnvGen.kr(env, 1, 24, 60, 20, 2).poll.midicps.poll,
8          Impulse.kr(6)))
9  }.play
```

The output range of the envelope is now $[60, 84]$. The idea is to represent pitch rather than frequency: a two-octaves interval starting at middle C (60 in MIDI notation) is converted to cycle-per-seconds (that is, a frequency in Hz) by invoking the midicps method. Generally speaking any algebraic operation can be applied to the output of any UGen, e.g. squared or abs, to mention a couple that we have already seen elsewhere. The poll method, which is encountered two times in the above example, is, roughly speaking, the server-side equivalent of postln. The need for such a specialized poll method is related with the client/server architecture of SuperCollider: even if audio synthesis processes are controlled by the client, they are implemented on the server. The former merely instructs the latter which is responsible for actually carrying out the instruction. So, how can the client probe the internals of the server? How can a user check for potential implementation errors? Audio monitoring is not always sufficient to debug some particular algorithm. Of course, it could happen that some particular errors are audible (it could be that they sound interesting, too); it is not recommended to debug in serendipity, however. It is important to also remark that scoping and plotting are not analytic tools —that is, they cannot show information at the level of audio samples. On the contrary, poll instructs the server to send a message to the client with the values of individual audio samples and at a given rate; the client, then, automatically prints these values to the post window. This way it is possible to monitor the output of some

UGen with the desired degree of accuracy. (Note that poll is applicable to both ar and kr signals.) For example:

```
1 {SinOsc.ar(Line.ar(50, 10000, 10).poll).poll}.play ;
```

produces

```
 1 Synth("temp__1198652111" : 1001)
 2 UGen(Line): 50.0226
 3 UGen(SinOsc): 0.00712373
 4 UGen(Line): 149.523
 5 UGen(SinOsc): -0.142406
 6 UGen(Line): 249.023
 7 UGen(SinOsc): -0.570459
 8 UGen(Line): 348.523
 9 UGen(SinOsc): -0.982863
10 UGen(Line): 448.023
11 UGen(SinOsc): -0.616042
12 UGen(Line): 547.523
13 UGen(SinOsc): 0.676455
```

The output of each UGen can now be printed to the post window. The SinOsc object oscillates in the $[-1, 1]$ range, while the Line UGen linearly progress from 50 to $10,000$.

Going back to the original SynthDef above, poll.round.poll.midicps.poll prints the EnvGen's output before and after its rounding, as well as after its conversion to Hz. Notice that in the post window, midi notes are expressed as float numbers. Indeed, midi conversion within SuperCollider can handle any note value and not just the "tempered" ones. Consider the following example:

```
1  {
2      var levels, times, env ;
3      levels = Array.fill(50, { arg x ; (x%7)+(x%3) }).normalize ;
4      times = Array.fill(49, 1).normalizeSum ;
5      env = Env.new(levels, times) ;

7      Pulse.ar(Latch.kr(EnvGen.kr(env, 1, 24, 60, 20, 2)
8          .poll.round.poll.midicps.poll,
9          Impulse.kr(6)))
10  }.play
```

Here, a continuous signal is converted to a discreet one in two different respects: on one hand, Latch generates 6 notes per second and, on the other hand, all pitches are quantized so that only integers within the [60, 84] range are possible. This acts as a digitalization process, the depth of which is defined by round's argument; a value of 0.5 in the example above, for instance, would quantize all pitches to quarter tones.

Figure **6.2** shows a part of the resulting UGen graph (Env's structure is rather large, since there are 50 breakpoints). Note the existence of Poll UGens, their internal trigger —an Impulse UGen with its frequency set to 10— and the operators Round and MIDICPS.

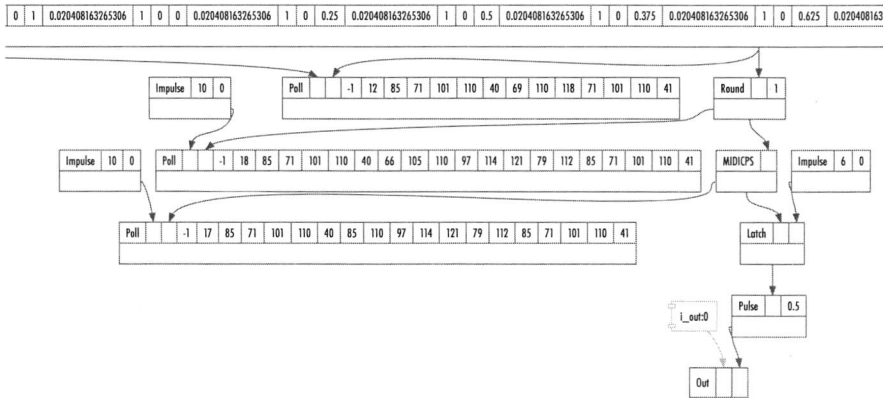

Figure 6.2   UGen graph of the SynthDef.

As clearly illustrated in Figure **6.2**, mathematical operators within a UGen graph are also UGens themselves. In particular, they are instances of `BinaryOpUGen` or of `UnaryOpUGen`. You do not have to worry about their meaning; it is important, nevertheless, to remark that all scsynth can understand and operate on within a synth is UGens. Consider the following code snippet:

```
1  {SinOsc.ar(200).round(0.5)}.play ;
2  {SinOsc.ar(200).abs-0.5}.play ;
```

The resulting graphs and signals are illustrate in Figure **6.3**.

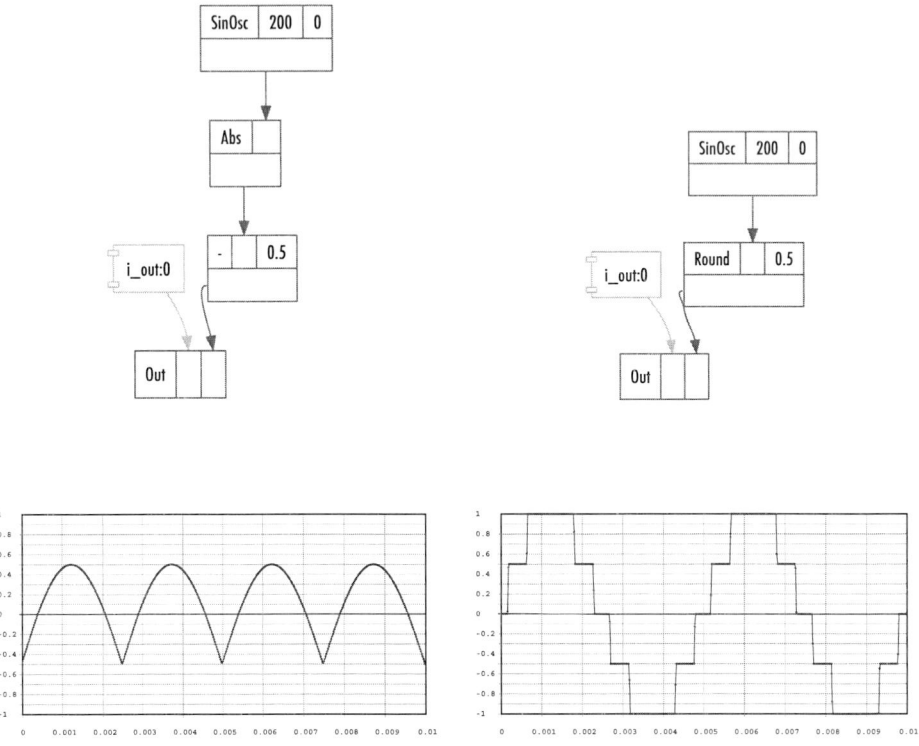

Figure 6.3 OpUGen: graphs and signals.

6.3 Sinusoids & sinusoids

It has been already been discussed that we can generate more "complex", "natural", or otherwise "interesting" signals by means of envelopes that control the amplitude or other synthesis parameters. Envelopes are (typically) unipolar signals, yet it is possible to also use bipolar signals in order to modulate synthesis parameters. Sinusoidal waveforms, for example, are more than just the means to produce the simplest of all sounds (a tone at a given frequency); they are also functions that describe regular variation around some equilibrium. Consider the following cases:

```
1  // minimal tremolo
2  { SinOsc.ar(mul: 0.5+SinOsc.kr(5, mul: 0.1)) }.play ;
3  // minimal vibrato
4  { SinOsc.ar(freq: 440+SinOsc.kr(5, mul: 5)) }.play ;

6  // with MouseX/Y
7  // tremolo
8  { SinOsc.ar(mul: 0.5 + SinOsc.kr(
9              freq: MouseX.kr(0, 10),
10             mul: MouseY.kr(0.0, 0.5))) }.play ;
11 // vibrato
12 { SinOsc.ar(freq: 440 + SinOsc.kr(
13             freq: MouseX.kr(0, 10),
14             mul: MouseY.kr(0, 10))) }.play ;
```

- **tremolo**: in music, a tremolo is a periodic variation of the perceived dynamics. Implementing a tremolo is actually very simple to synthesize. It suffices that we add to the amplitude of some oscillator (or of some other carrier signal) the output of another oscillator (or of some other modulating signal). This way, the carrier's amplitude will vary periodically according to the modulating oscillator's frequency and with respect to the latter's amplitude offset. In the example above, the original amplitude is 0.5 (the value of mul), to which a modulating signal that oscillates 5 times per second between -0.1 and 0.1 is added. The result is a sine tone of an amplitude that periodically varies within the $[0.4, 0.6]$ range. In lines 8–10 the mouse is used as the means to interact with the two parameters of the tremolo. With such techniques it is possible to emulate the characteristic qualities of certain real-world instruments that use this technique, e.g. the wind ones.

- **vibrato**: the same modulation technique applied to frequency results in a vibrato effect. Here, the modulating oscillator controls the variation of the carrier's frequency. Assuming that f_1, amp_1, f_2, amp_2 are, respectively, the frequencies/amplitudes of the audio and control oscillators, then the carrier's frequency (f_1, constant so far) would periodically vary (with respect to f_2) between $f_1 - amp_2$ and $f_1 + amp_2$. Remember that the output of the modulating oscillator varies in the $\pm amp_2$ (that is $[-5, 5]$ herein) range, so that when added to f_1 (that is 440 herein) a sine tone with a frequency that periodically (and at a speed of 5 times per second) oscillates between 435 and

445 will be produced. An example with mouse interaction is also given in the code above. Vibrati are often encountered in a musical context; consider e.g. a violinist who slightly, but continuously, slides their finger around a particular stop. Note also that both tremolo and vibrato are standard techniques in operatic singing.

The following example summarizes the above:

```
1  SynthDef("tremVibr",
2          { arg      freq = 440, mul = 0.15,
3                  tremoloFreq = 5 , tremoloMulPercent = 5,
4                  vibratoFreq = 10, vibratoMulPercent = 5   ;
5          var tremoloMul = mul*tremoloMulPercent*0.01 ;
6          var vibratoMul = freq*vibratoMulPercent*0.01 ;
7          var tremolo = SinOsc.kr(tremoloFreq, 0, tremoloMul) ;
8          var vibrato = SinOsc.kr(vibratoFreq, 0, vibratoMul) ;
9          var sinOsc = SinOsc.ar(freq+vibrato, 0, mul+tremolo) ;
10         Out.ar(0, sinOsc) ;

12 }).add ;
```

In lines 8 and 9, tremolo and vibrato are defined as two sinusoidal oscillators with frequency tremoloFreq and vibratoFreq, respectively. The "depths" of the tremolo and the vibrato effects are expressed proportionally: In line 5, tremoloMulPercent is first converted to a percentage (being multiplied with 0.01) and then multiplied with mul, so that the tremolo depth coefficient tremoloMul is computed. If mul = 0.5 and tremoloMulPercent = 10, then tremoloMul will be 10 % of mul, that is 0.05. Therefore, the signal would have a tremolo effect within the $[-0.05, 0.05]$ range and its amplitude would fluctuate between $[0.45, 0.55]$. vibratoMul is computed in a similar way. The sinOsc variable (note that this is just the name of a variable and should not be confused with the SinOsc UGen) holds the resulting signal which is, eventually, passed to the Out UGen so that audio output is generated in our speakers.

Given such a SynthDef, it is possible to build a GUI in order to control its arguments. The idea is to have sliders, labels and value indicators for every parameter. As we are dealing with just a few parameters here, we can "manually" define all individual components as needed. Another scenario would be to algorithmically generate the GUI elements. In both cases we would rely on the following GUI classes:

1. Slider: its semantics are rather straightforward: a parent window (an instance of Window in the example code below[1]) and a rectangle (an instance of Rect) that defines the Slider's coordinates and dimension.
2. StaticText: a display-only text field. The syntax is similar: a parent Window and a Rect with the coordinates/dimensions are expected. The text string to be printed is defined using the string_ method.

Given an array with the names of the SynthDef's arguments, it is then possible to:

1. algorithmically generate the graphical objects
2. algorithmically define the actions associated with them

The only complexity lies in properly scaling the sliders' values (all GUI controls out values in the $[0, 1]$ range) to the ones expected by each argument: frequency freq could vary in the $[50, 10000]$ range; amplitude mul in the $[0, 1]$ range; tremolo/vibrato rates are typically expected to be in a lower frequency (subsonic) region, e.g. $[0, 15]$. Their depths should be in the $[0, 100]$ range, since they are expressed as percentages. It is possible to scale the sliders' output within our SynthDef, yet this is considered bad practice —it breaks the rule of modularity and makes our SynthDef GUI-dependent. Generally speaking, it is best to keep data models and sound synthesis processes completely independent from GUI elements. This way, it is possible to easily switch between different interfaces without having to change our original models (the SynthDef in this case). A SynthDef cannot impose any particular constraints (invariables) to a GUI, so there cannot be any guarantee that the latter's components will have their values properly scaled to the expected ranges. In the following example this problem is addressed by means of an IdentityDictionary —a data structure that associates a unique key to a value, much the same way that an ordinary dictionary associates a lemma with a definition— which binds arguments with their expected ranges ([minval, maxval]). In particular, there are three arrays containing the labels, the control sliders and the value fields for all arguments (lines 14–24). The functionality of the GUI is defined within a loop that iterates through all available elements in controlDict. value indicates the expected range, thus the key name may be retrieved by calling the controlDict.findKeyForValue(value) method —e.g. if value is [50, 1000],

---

[1] Note that window's height is calculated by adding 30×2 pixels to the height of the tallest element.

the associated key name would be "freq". Then, the slider's value (originally
in the $[0, 1]$ range) is appropriately scaled by means of a simple multiplication
with the $range_{max} - range_{min}$ difference ($1000 - 50$ in this case) and the addition of the $range_{min}$ (here: 50) to create the lower offset. We can, then, use
labelArr[index].string = ... and the slidArr[index].action = ... to assign
each label with the name for an argument and each slider with the corresponding synth's parameter. Within the slider's action, valueArr[index].string is
also used so that the value of the value field and that of the slider are identical.
In detail, each slider's action performs four tasks (lines 32–35):

1. it scales the value of the slider so that it matches the expected range (as
   explained above).
2. it prints the name of the corresponding synthesis parameter and its value to
   the post window.
3. it sets the synth's name parameter to paramValue.
4. it, finally, updates the textField index in valueArr to show the paramValue.

```
1   var aSynth =  Synth.new("tremVibr");
2   var controlDict = IdentityDictionary[
3       "freq"                  ->      [50, 1000],
4       "mul"                   ->      [0,1],
5       "tremoloFreq"           ->      [0, 15],
6       "tremoloMulPercent"     ->      [0, 50],
7       "vibratoFreq"           ->      [0, 15],
8       "vibratoMulPercent"     ->      [0, 50]
9   ];

11  var window = Window.new("Vibrato + tremolo",
12      Rect(30,30, 900, controlDict.size+2*30)) ;

14  var labelArr = Array.fill(controlDict.size, { arg index ;
15      StaticText( window, Rect( 20, index+1*30, 200, 30 )).string_( 0 ) ;
16  }) ;

18  var slidArr = Array.fill(controlDict.size,
19      { |index| Slider( window, Rect( 240, index+1*30, 340, 30 )) }) ;

21  var valueArr = Array.fill(controlDict.size,
22      { |index| StaticText( window, Rect( 600, index+1*30, 200, 30 ))
23          .string_( 0 ) ;
24  }) ;

26  controlDict.do({ arg value, index ;
27      var name = controlDict.findKeyForValue(value) ;
28      var range = value[1]-value[0] ;
29      var offset = value[0] ;
30      labelArr[index].string_(name) ;
31      slidArr[index].action = { arg theSlid ;
32          var paramValue = theSlid.value*range + offset ;
33          [name, paramValue].postln ;
34          aSynth.set(name, paramValue) ;
35          valueArr[index].string_(paramValue.trunc(0.001) ) ;
36      }
37  }) ;

39  window.front ;
```

6.4 Pseudo-random signals

The idea of controlling an oscillator with another can be generalized to account
for all sorts of UGens —both as carriers and modulators. The time-profile of
the control signal depends on the kind of UGen used. The following examples demonstrate how we can use other kinds of UGens to control the vibrato
of a sinusoid. The scope method is used to illustrate the waveform profile of
each control signal (frequency has been changed to 1000 Hz so that the latter is
clearly audible). Note that typically those UGens receive a LF prefix (shorthand
for Low Frequency), as they are meant to be primarily used as slower control
signals.

```
1 { SinOsc.ar(1000) }.scope ;
2 { SinOsc.ar(freq: 440+SinOsc.kr(2, mul: 50), mul: 0.5) }.play ;

4 { LFSaw.ar(1000) }.scope ;
5 { SinOsc.ar(freq: 440 + LFSaw.kr(2, mul: 50), mul: 0.5) }.play ;

7 { LFNoise0.ar(1000) }.scope ;
8 { SinOsc.ar(freq: 440 + LFNoise0.kr(2, mul: 50),mul: 0.5) }.play ;

10 { LFNoise1.ar(1000) }.scope ;
11 { SinOsc.ar(freq: 440 + LFNoise1.kr(2, mul: 50), mul: 0.5) }.play ;
```

The difference between the sinusoidal and the saw-tooth oscillators is strik-
ing. The other two UGens generate pseudo-random signals. LFNoise0, in particular, generates amplitude values that are sustained until a new value is calculated: notice the step pattern in the scope window. In the same vein, LFNoise1
generates a pseudo-random signal at a given frequency, yet this time the output
values are linearly interpolated. In other words, there are no "steps" between a
value and the next, but, instead, a gradual transition. The behavior of the two
UGens is clearly illustrated in the following piece of code:

```
1  {SinOsc.ar(LFNoise1.ar(10, mul:200, add: 400))}.play ;
2  {SinOsc.ar(LFNoise0.ar(10, mul:200, add: 400))}.play ;
```

In both cases the frequency of the oscillator is controlled by a random number generator that updates its state 10 times per second. Following `mul` and `add` arguments, the control signal varies randomly in the range of $[-1, 1] \times 200 + 400 = [200, 600]$. Then, `LFNoise0` causes the oscillator to jump to different frequencies while `LFNoise1` causes continuous linear transitions (glissandi) between random frequencies. `LFNoise0` can be thought of as a "sampled and held" version of `LFNoise1`: as if sampling the latter's output, holding the value for a cycle, sampling again, etc. It turns out that this is exactly how `Latch` behaves, as demonstated in the following code snippet:

```
1  { SinOsc.ar(LFNoise0.kr(9, 400, 500), 4, 0.2)}.play ;

3  // the same, but less efficient
4  { SinOsc.ar(Latch.ar(LFNoise1.ar, Impulse.ar(9)) * 400 + 500, 4, 0.2) }.play;
```

`LFNoise0` can be also be used as (in a way) a sequencer, i.e. as a train of unique discrete values. Consider the following example, implementing two "improvisers":

```
 1  SynthDef(\chromaticImproviser , { arg freq = 10 ;
 2      Out.ar(0, SinOsc.ar(
 3          freq:      LFNoise0.kr(freq, mul:15, add: 60).round.midicps,
 4          mul:
 5          EnvGen.kr(Env.perc(0.05), gate: Impulse.kr(freq), doneAction:2)
 6      )
 7  }).play ;

 9  SynthDef(\modalImproviser , { arg freq = 10;
10      var scale = [0, 0, 0, 0, 3, 3, 4, 5, 5, 6, 6, 7, 7, 7, 10]+60 ;
11      var mode = scale.addAll(scale+12).midicps ;
12      var range = (mode.size*0.5).asInteger ;
13      Out.ar(0, SinOsc.ar(
14          freq:      Select.kr(LFNoise0.kr(freq,
15              mul: range,
16              add: range).round, mode),
17          mul:
18          EnvGen.kr(Env.perc(0.05), gate: Impulse.kr(freq), doneAction:2)
19      )
20  }).play ;
```

Both the "improvisers" generate random sequences of 10 notes per second at a given register. In both cases the audio generator is a sinusoidal oscillator wrapped with a percussive envelope. (Note that, as shown in the example, if message play is directly sent to a SynthDef, the server would automatically create an instance of Synth once the SynthDef is created.) The first SynthDef generates a sequence of chromatic intervals between [45, 75] in midi notation. Midi notes are nothing but an indication of pitch, much in the same way a piano's white and black keys are. In midi notation the piano's "middle C" is 60; therefore the first improviser's output varies from the 15th key before the middle C (that is an A, 2 octaves lower) to the one the 15th after it (that is a E♭, 2 octaves up). The frequency of the oscillator is the LFNoise0's output multiplied by 15, with 60 added.

The second SynthDef calculates pitches differently. A scale is defined as a given sequence of steps (each step is a semitone) starting at 0. Then, 60 is added to the intervals so that all pitches are transposed to the desired octave. In this way, a particular trajectory of pitches starting from the middle C (60) and ending at the next B♭ (70) is defined. The resulting midi notes are the following:

```
1  [0, 0, 0, 0, 3, 3, 4, 5, 5, 6, 6, 7, 7, 7, 10]+60
2  [ 60, 60, 60, 60, 63, 63, 64, 65, 65, 66, 66, 67, 67, 67, 70 ]
```

or C, E♭, E, F, F♯, G, B♭, some of which repeated. The actual sequence is given by the scale, extended with a copy of itself transposed an octave up. Remember that an octave interval in midi notation is represented as an increase/decrease of 12. The mode array is then converted to Hz by midicps. Frequency is determined by a Select UGen; Select's *ar and *kr methods expect two arguments, which and array: given an array of signals, Select returns the one indexed by which. In the above example, the idea is to randomly Select a frequency from the mode array using LFNoise0. The latter generates a number in the $[-r, r]+r = [0, r \times 2]$ range, where the variable r (here range) is defined as half of the size of mode. Therefore, if mode.size = 26, then range = 13, and LFNoise0 will fluctuate in the $[0, 26]$ range. Since the output values from LFNoise0 are used as indices, they must be integer values, hence the use of round. It has been noted that mode contains the same values more than once. This is a primitive, albeit effective, way to raise the probability of certain pitches. In this way, some steps in the scale are emphasized as they are more likely to be selected: the fundamental and, then, the minor third, the fourth, the augmented fourth and the fifth —thus achieving a somehow "bluesy" quality. Note that the oscillator's output should be enveloped each time there is a change in pitch, or in other words, the oscillator should be "in sync" with LFNoise0. Accordingly, the gate argument is driven by an Impulse which triggers the envelope at the same frequency as that of LFNoise0.

Another UGen typically used to produce triggers is the pseudo-random pulse-generator Dust. Dust produces a train of pulses in the $[0, 1]$ range, but, unlike Impulse, these pulses are stochastically distributed in time. The average number of pulses per second is given by the first argument of the UGen (density). The arguments and their default values for both *ar and *kr methods are: density = 0.0, mul = 1.0, add = 0.0. If we replace Impulse with Dust in the previous example, leaving the rest of the code unchanged, we would still get more or less the same number of triggers per second, yet not equally distributed in time. Consider the next example where the mouse is used to control the density:

```
1 { Dust.ar(MouseX.kr(1, 500)) }.play ;
```

Using Dust as a trigger is fairly common in SuperCollider. The following example is taken from the help file of another interesting UGen, namely TRand. TRand generates a pseudo-random value in a range defined by its first two arguments lo = 0.0 and hi = 1.0. Each time a trigger is registered in the third argument, a different value is generated at the UGen's output. In the following example, the first part generates a trigger Impulse so that at each new pulse, TRand modulates the frequency of SinOsc. This is yet another way to implement a *Sample & Hold* technique. Replacing Impulse with Dust will result in a series of randomly distributed pulses (and therefore of envelopes) having the average density set by freq.

```
 1 // deterministic
 2 {
 3     var trig = Impulse.kr(9);
 4     SinOsc.ar(
 5         TRand.kr(100, 900, trig)
 6     ) * 0.1
 7 }.play;

 9 // stochastic
10 {
11     var trig = Dust.kr(9);
12     SinOsc.ar(
13         TRand.kr(100, 900, trig)
14     ) * 0.1
15 }.play;
```

6.5 Busses

It is now time to introduce busses. The bus concept derives from analog audio. As the Figure **5.4** illustrates, a helpful metaphor is that of a pipe where signal flows. Each Bus is associated with a unique index (a numerical identifier) which can be used to refer to it. Both control and audio busses are available in Super-Collider. The latter are intrinsic to the server in that the signals they carry are both produced and consumed by the former. As far as audio rate busses are concerned, however, we both encounter ones routing audio to other elements within the Server as well as ones routing audio to the outside world —e.g. to the drivers of some external audio interface, or from some external input (a microphone). In relation to audio busses, the server does not differentiate between busses used internally, i.e. "private", and busses used to connect with external audio interfaces, i.e. "public" busses— both are assigned an index and their structure is identical in all respects. There is an important convention, however: the first n audio busses $(0, ..., n)$ are reserved for the outputs and inputs of the computer's audio interface; the rest may then be used internally. The exact number of public and private numbers depends on the type of soundcard in use as well as on the SuperCollider settings. The standard configuration assumes a stereo output at busses 0 and 1 and a stereo input at busses 2 and 3. By default, from 4 forward, the busses can be used internally.

The following example illustrates how a simple `SynthDef`'s output can be routed to more than one audio bus using the `Out` UGen. `Out`'s arguments indicate the index of the Bus to write to and the signal to be written. In line 2 of the code below we open a scope window to visually monitor the first 4 audio busses (indices $0, 1, 2, 3$). Bus 0 corresponds to the left channel of a classic stereo configuration. In line 4 the signal is moved to the right channel (with index 1). Then, in lines 5 and 6, the signal is routed to the busses 2 and 3, respectively. Of course, with a stereophonic interface it is impossible to hear the sound in these busses (we can still visually inspect them, however). The signal is, eventually, moved back to the left channel (line 7).

```
1 SynthDef(\chan , {arg out; Out.ar(out, LFPulse.ar(5)*SinOsc.ar(400))}).add ;
2 s.scope(4) ; // monitor for busses
3 x = Synth(\chan ) ;
4 x.set(\out , 1) ; // on the right
5 x.set(\out , 2) ; // it's here but private
6 x.set(\out , 3) ; // same as before
7 x.set(\out , 0) ; // on the left
```

We can have multiple Outs (even operating at different rates) within the same SynthDef. The following code writes two different signals in two adjacent busses. Then, by varying out, we can "move" this arrangement to another set of consecutive busses.

```
1  SynthDef(\chan , { arg out = 0;
2      Out.ar(out, LFPulse.ar(5)*SinOsc.ar(400)) ;
3      Out.ar(out+1, LFPulse.ar(7)*SinOsc.ar(600)) ;
4  }).add ;
5  s.scope(4) ; // monitor for busses
6  x = Synth(\chan ) ;
7  x.set(\out , 1) ; // to right and to private
8  x.set(\out , 0) ; // from start
```

Managing multichannel systems is therefore straightforward: e.g. we can use busses 0...7 to drive a sound card with 8 audio outputs, etc. Out can be used to control the routing of the various signals. In reality, things are even simpler, as to be shown briefly. A possible situation is outlined in Figure **6.4**, in relation to different possible hardware configurations.

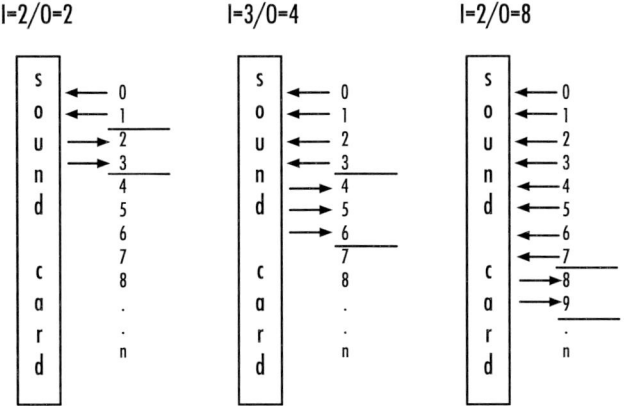

Figure 6.4   Private and public busses in various hardware configurations.

In the following example a stethoscope is used to visualize the contents of busses 0...3: given a standard configuration these would be the stereo output and the stereo inputs.

```
1 s.scope(4,0) ;
2 {Out.ar(0, In.ar(2))}.play ;
```

Then, in line 2:

1. the UGen In is introduced, which is Out's equivalent and is used for reading audio signal from a bus;
2. In reads from Bus 2, which is assumed to be one of the soundcard's audio inputs. The signal is now available within at scsynth;
3. using Out, the input signal is routed to the bus 0, which corresponds to the left output of a standard stereo soundcard.

(Beware of possible feedback if a microphone close to the speakers is connected in bus 2). The situation is shown in Figure **6.5**.

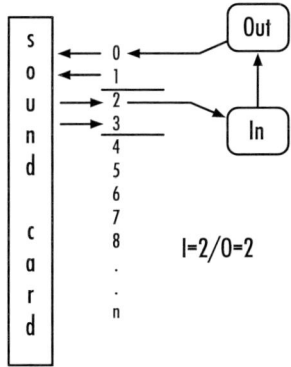

Figure 6.5   Routing among input and output busses.

Now suppose that we use a quadraphonic audio interface, such as the second in Figure **6.4**. In this case, bus 2 represents an output, rather than an input. This can be very confusing: identical code could have dramatically different semantics, depending on the audio interface used —this would be unacceptable. Hopefully there is a dedicated UGen we can use to read a soundcard's inputs irrespective of their indices, namely SoundIn. SoundIn's first argument is indeed an index, yet it will be automatically biased so that it always represents an input bus. Therefore, in=0 represents "the first available audio input", rather than the bus 0. In the next example, the first available audio input (typically the computer's built-in microphone) is routed to the output bus 0:

```
1  SynthDef(\in , {Out.ar(0, SoundIn.ar(0))}).add ;
2  x = Synth(\in ) ;
```

On a typical computer with a stereo sound card, index 0 would represent bus 2. However, given an audio interface with 10 outputs, it will represent bus 10. With SoundIn we can, therefore, use relative, rather than absolute, indices which results in safer, portable and more efficient code.

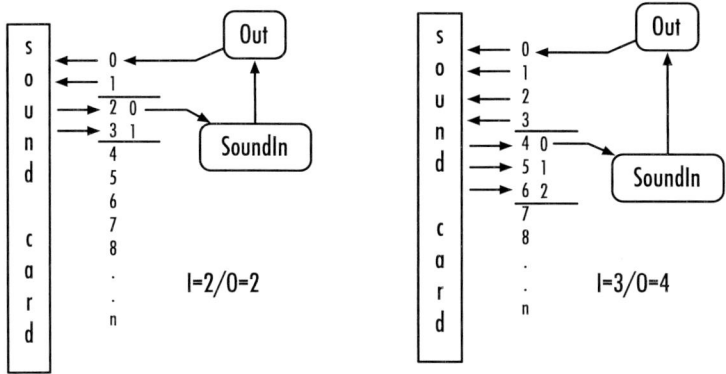

Figure 6.6 Semantics of SoundIn.

While public busses are an interface to the outside world, private ones allow for internal routing configurations of arbitrary complexity. They are particularly useful in emulating classic analog audio synthesis: e.g. having several synths that may be connected/disconnected at will. In this way we can create more complex signals. Consider also delay lines and reverb. In space, each reflection causes an acoustic wave to be delayed so that live sound is always a mixture of several delayed versions of some original sonic event. A minimal delay unit would merely generate a copy of the input signal delayed for some small amount of time. A series of such units could be, then, used to simulate the effect of a reverberation. There are many ways to implement such units: in

any case, SuperCollider already features an implementation of FreeVerb —a reverb widely used in open-source communities. The following code comprises
two synths: the first is a 50 ms pulse multiplied with a Line— a linear-ramp
envelope which, in this case, generates a constant 1 for the given duration and
then frees the wrapping synth; the second applies FreeVerb to the former:

```
1  {Pulse.ar(100)*Line.kr(1,1, 0.05, doneAction:2)}.play ;
2  {FreeVerb.ar(Pulse.ar(100)*Line.kr(1,1, 0.05, doneAction:2))}.play ;
```

There appears to be no audible difference between the original and its reverberated version. Line de-allocates the entire synth once the pulse is over and,
therefore, the reverberation never happens. The problem can be easily solved
using Bus.

```
1  (
2  SynthDef(\blip , {arg out = 0;
3      Out.ar(out, Pulse.ar(100)*Line.kr(1,1, 0.05, doneAction:2))
4  }).add ;

6  SynthDef(\rev , {arg in, out = 0;
7      Out.ar(out, FreeVerb.ar(In.ar(in)))
8  }).add ;
9  )

11 Synth(\blip , [\out , 0]) ;

13 //vs
14 (
15 ~revBus = Bus.audio(s,1) ;
16 ~rev = Synth(\rev ) ;
17 ~rev.set(\in , ~revBus) ;
18 Synth(\blip , [\out , ~revBus]) ;
19 )

21 x = {Out.ar(~revBus, Dust.ar(3))}.play ;
22 x.free ;
```

The first block of code (lines 1–9) defines two SynthDefs, the first being the above short pulse generator and the second being a minimal reverberation module —Figure **6.6** demonstrates the latter's UGen Graph.

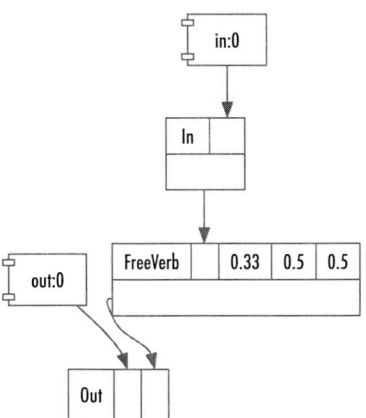

Figure 6.7   A minimal reverb module.

The synth in line 11 generates an impulse without any reverberation. Following the code block, however, the output of the blip synth is routed to the reverberation module using a private Bus object. Note that we can directly pass a Bus object as an argument to whatever UGen expects a bus index. Bus' audio method is a constructor for an audio rate bus —a 1-channel bus allocated on server s, in this case— as specified by the two arguments. Even if the server only understands monophonic busses, Bus can be used for the construction of multichannel ones: the Bus object will allocate and handle a series of busses with adjacent indices, to be used as a group. As far as the original problem is concerned, in the above example the synth ~rev which reads the bus ~revBus keeps producing output even after the original source —that is, the synth blip— is automatically de-allocated. Try evaluating the code in line 18 again; a new blip sound is sent through the reverberation unit which is still active. As shown in line 21, we can also send any other kind of signal to revBus. Another benefit comes from computational savings. Reverb units, for example, use a lot of processing power. If each note had its own reverb, your computer would quickly start to glitch from the load. But with a Bus, a single Reverb can be used with all of your other note events.

When using busses, however, we have to consider the so-called order of execution of the synths. In the server, all nodes (that is, groups and synths) are arranged in a tree that determines the order they should be executed. For each sampling period $T = \frac{1}{sr}$, all synths' graphs are scanned so that the output values for each one are computed. Yet, it is according to the execution graph that the server decides the output of which graphs to compute first. When using busses, we must ensure that signal we want to read off of a bus by other synths are written to the busses first. Consider the following code:

```
1  (
2  SynthDef(\pulse , {arg out = 0, freq = 4;
3       Out.ar(out, Pulse.ar(freq, 0.05))})).add ;
4  )

6  // 1.
7  ~revBus = Bus.audio(s) ;
8  ~src = Synth(\pulse , [\out , ~revBus]) ;
9  ~rev = Synth(\rev ) ;
10 ~rev.set(\in , ~revBus) ;

12 // 2.
13 ~revBus = Bus.audio(s) ;
14 ~rev = Synth(\rev ) ;
15 ~src = Synth(\pulse , [\out , ~revBus]) ;
16 ~rev.set(\in , ~revBus) ;

18 // 3.
19 ~revBus = Bus.audio(s) ;
20 ~src = Synth(\pulse , [\out , ~revBus]) ;
21 ~rev = Synth(\rev , addAction:\addToTail ) ;
22 ~rev.set(\in , ~revBus) ;

25 // 4.
26 ~revBus = Bus.audio(s) ;
27 ~src = Synth(\pulse , [\out , ~revBus]) ;
28 ~rev = Synth.after(~src, \rev , addAction:\addToTail ) ;
29 ~rev.set(\in , ~revBus) ;
```

In the first case, ~src is allocated before ~rev. This results in no sound. Whenever a new synth is allocated, it is put by default in the head of the execution graph, so that its output is the first to be computed. We would expect the last allocated synth to be the one computed last, but this is not the case in the server. The execution graph can be visualized in two ways, as shown in the code below:

```
1  s.queryAllNodes ;
2  s.plotTree ;
```

The first method prints the graph on the post window; the latter rather builds a GUI displaying the various synths/groups and their order. In the case of the above example concerning reverberation, the resulting graph would look like this:

```
1  localhost
2  NODE TREE Group 0
3      1 group
4          1001 rev
5          1000 dust
```

There is a default group, here including two synths. The execution order is ~rev→ ~pulse. As shown, the reverb is calculated *before* the impulse signal, so that the former will simply produce 0 at its output, given that there is no input to process. The situation is illustrated in Figure **6.8**.

The order of execution in the case of the second code block is exactly the opposite: the reverberation unit is created before the pulse source, so that the latter is computed first: therefore, there is a value for the reverb synth to process. In the third code block, an addAction is used to explicitly place the reverberation synth at the end (addToTail) of the execution graph —addToHead can be used if the opposite result is desired. Finally, in the fourth code block, the synth is created using the Synth's after, rather than the new, constructor, as the method enables us to place the resulting synth after a particular node (determined by the first argument) —and indeed a before method does exists, that allows to place a synth before another one in the execution graph.

As already mentioned, busses can be also used to read/write control signals. In the following example, the pseudo-UGens MouseX and MouseY are used to manage two synths (~tri1 and ~tri2), that substantially wrap triangular

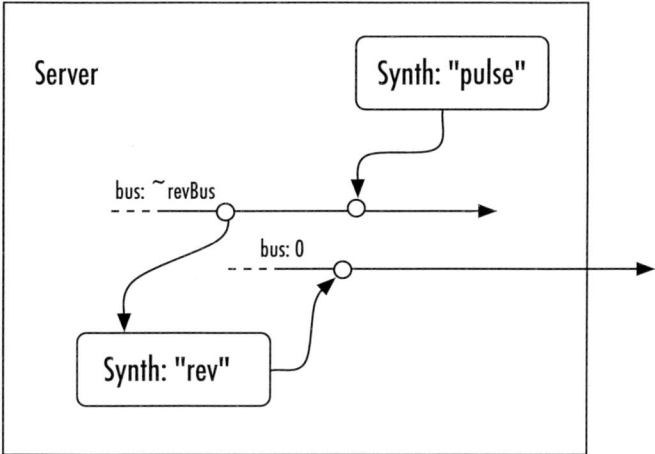

Figure 6.8   Wrong order of execution.

wave oscillators (notice that low-frequency oscillators can be also used at audio rate). Herein, the control interface is kept distinct from the audio generators. An important distinction between audio and control busses is that, for the control busses, the order of execution is not applicable. While audio signal on busses are zeroed out after each control cycle, control busses retain their value until updated directly. Accordingly, the ~tri1 and ~tri2 synths may both read the input from the bus on which ~mouse writes to. The code block between lines 18–23 shows how to use the get method as an interface to retrieve values from the control bus. The method get expects a callback function as an argument; the argument of the latter will be, in turn, given the bus' current value automatically.[2]. In this particular case, the callback function converts the value to a MIDI value, rounds it up and eventually posts it to the post window. The same process is repeated for the ~yBus. The last lines in the above piece of code demonstrate the set method: a new control bus ~vBus is allocated and used

---

[2] We need a function to access this value because it is stored server-side and, therefore, a series of OSC messages between the server and the client have to be exchanged before it is available to the latter. Operations involving OSC communication are asynchronous: it cannot be known in advance when exactly a message will arrive. Accordingly, the provided function will be evaluated once the desired value has indeed arrived, and the latter will be passed as an argument to the user-defined function.

to control the frequency of yet another "tri" synth, which writes audio to the right channel output (as shown in the scope window invoked at line 27); in line 26 we "get" the values from ~xBus, which are, then, transposed an octave up and written to ~vBus. This way, we get a copy of the ~tri1 synth in the right channel, only that the latter is transposed an octave higher.

```
1  (
2  SynthDef(\tri , {arg out = 0, inFreq = 0, amp = 1;
3      Out.ar(out, LFTri.ar(In.kr(inFreq), mul:amp))}).add ;

5  SynthDef(\mouse , {arg out1, out2 ;
6      Out.kr(out1, MouseX.kr(36, 96).round.midicps);
7      Out.kr(out2, MouseY.kr(36, 96).round.midicps);
8  }).add ;
9  )

11 ~xBus = Bus.control(s) ;
12 ~yBus = Bus.control(s) ;
13 ~mouse = Synth(\mouse , [\out1 , ~xBus, \out2 , ~yBus]) ;
14 ~tri1 = Synth(\tri , [\amp , 0.5]) ;
15 ~tri2 = Synth(\tri , [\amp , 0.5]) ;
16 ~tri1.set(\inFreq , ~xBus) ;
17 ~tri2.set(\inFreq , ~yBus) ;
18 (
19 ~xBus.get{|v|
20     v.cpsmidi.round.postln;
21     ~yBus.get{|v| v.cpsmidi.round.postln};
22 } ;
23 )
24 ~vBus = Bus.control(s) ;
25 ~tri3 = Synth(\tri , [\inFreq , ~vBus, \out , 1]) ;
26 ~xBus.get{|v| ~vBus.set(v*2); }
27 s.scope ;
```

Note that while control busses are very convenient, they typically require the presence of one or more In UGens. This means that the corresponding SynthDefs are necessarily context-dependent, since they assume the presence of particular signals in their control inputs. An interesting alternative is the use of the map method.

```
1  (
2  SynthDef(\tri , {arg out = 0, freq = 440;
3      Out.ar(out, LFTri.ar(freq))}).add ;

5  SynthDef(\mouse , {arg out ;
6      Out.kr(out, MouseX.kr(20, 5000, 1));
7  }).add ;

9  SynthDef(\sine , {arg out, freq = 10 ;
10     Out.kr(out, SinOsc.kr(freq, mul: 200, add:500));
11 }).add ;
12 )

14 ~kBus = Bus.control(s) ;
15 ~mouse = Synth(\mouse , [\out , ~kBus]) ;
16 ~tri1 = Synth(\tri ) ;
17 ~tri1.map(\freq , ~kBus) ; // from mouse
18 ~mouse.run(false) ; // las value from mouse
19 ~mouse.run ; // again
20 ~sine = Synth(\sine , [\out , ~kBus]) ; // from sine, overwritten
21 ~sine.run(false) ;
22 ~tri1.set(\freq , 100) ; // fixed
23 ~tri1.map(\freq , ~kBus) ; // from mouse
24 ~sine.run ; // from sine
```

In this example there are three SynthDefs, one is meant for the synthesis of audio signals ("tri") and the remaining two for generating control ones. After the allocation of a control bus (line 14), a control synth is defined (line 15) that writes the mouse output to ~kBus. Then, an audio synth is created (line 16), the frequency of which (freq) is mapped (using map) onto ~kBus. This way, the mouse directly controls the frequency of the audio synth without including an explicit In UGen or invoking the get method. Notice that when mouse interaction stops (line 18), the bus keeps on holding the last generated value. On line 19, the mouse synth resumes and on line 20 a different control signal (~sine) is routed on the same bus; the latter now overwrites the mouse synth output and controls the frequency. When ~sine is paused (line 21) the mouse's output becomes once again available. When a value is passed to the synth's argument, however (line 22) the connection with the bus is lost. On line 23 the mapping is re-established, but when ~sine resumes (line 24), the output of the mouse synth is, once again, overwritten. The map method makes can make the SynthDef independent of any interface and/or control I/O schemata.

It should be noted that audio and control signals behave differently when more signals are simultaneously written into them. As already discussed, in the case of control busses, the last signal overwrites any previous ones. In the case of audio busses, however, signals can be treated in a couple of different ways. When routing more than one signal with Out.ar(0, ...), the signal that exists in the bus and the new signal are summed together. It makes perfect sense since we do want all these signals to "mix". Consider the following code:

```
1  (
2  SynthDef(\tri , {arg out = 0;
3      Out.ar(out, LFTri.ar)}).add ;

5  SynthDef(\sin , {arg out ;
6      Out.ar(out, SinOsc.ar(1000));
7  }).add ;

9  SynthDef(\mix , {arg out = 0, in ;
10     Out.ar(out, In.ar(in));
11 }).add ;
12 )

14 b = Bus.audio ;
15 x = Synth(\tri , [\out , b]) ;
16 y = Synth(\sin , [\out , b]) ;
17 b.scope ; // monitoring
18 z = Synth.tail(nil, \mix , [\in , b]) ;
```

Here, a SynthDef mixes the inputs of all private busses (9-11), and then writes them to the sound card's output. The mixer synth (z) is allocated last, but the tail method ensures the proper order of execution. The two synths (x, y) write on the same bus that z reads from and both signals are mixed in bus b. As illustrated, it is possible to monitor a bus (both audio and control) using the scope method (line 17).

However, other UGens that write to an audio bus can interact with previous sound on the bus in different ways: without entering into details, ReplaceOut will zero out the bus before writing new signal while XOut accepts an xfade parameter that will mix the new and previously written signal.

6.6 Procedural structure of SynthDef

What is a SynthDef? It is the equivalent of a synthesizer's electronic schematics. SynthDefs are written in the SuperCollider language (sclang), yet, when they are sent to scsynth, they are compiled to an internally-used binary representation. In fact, scsynth knows nothing about sclang. Given the existence of an appropriate interpreter, SynthDefs could be written in any language, e.g. as long as they are eventually translated to the appropriate binary format, it makes no difference for the server. Therefore, a SynthDef is merely a description of a virtual synthesizers's structure —nevertheless, one that does take advantage of the expressiveness of the SuperCollider language. Consider the following code:

```
1  SynthDef(\chan , { arg out = 0 ;
2      8.do({|i|
3          Out.ar(out+i, LFPulse.ar(i+5)*SinOsc.ar(i+1*200))
4      }) ;
5  }).add ;
```

Here, the UGen graph is described iteratively. For each of the 8 iterations a pulse-modulated sinusoidal signal is created, its parameters being defined as a function of the counter and its output being routed on a series of consecutive busses for which the indexes of each are also given as a function of the counter. Using the following code, we can hear the first two signals (of course, assuming that a stereo sound card is available) and visually inspect all busses.

```
1  s.scope(8) ; // monitor for first 8 busses
2  x = Synth(\chan ) ;
3  x.set(\out , 1) ; // move to the right
4  x.set(\out , 0) ; // from start
```

The code is very compact, and also specifies a set of relationships that cannot be shown in the UGen graph. Consider the following code, too:

```
1  SynthDef(\chan  , { arg out = 0, freq = 200, kFreq = 5 ;
2      8.do({|i|
3          Out.ar(out+i, LFPulse.ar(i+kFreq)*SinOsc.ar(i+1*freq))
4      }) ;
5  }).add ;
```

The resulting graphs are shown in Figure **6.9**. Remember that figures of this type are obtained through automatic analysis of a SynthDef's structure and, therefore, what we see are the very same data structures accessible to the server itself. The expressive power of the SuperCollider language is well exemplified in these cases: with just a few lines of code we have created a very complex UGen graph.

On a typical stereo interface only the first two signals will be heard (busses 0 and 1). SuperCollider features a Mix UGen, however, that can be used to mix together all existent channels in the case of a multi-channel signal —the expected argument is an array of signals. The next code snippet is a modified version of the example in discussion; Mix sums together all 8 signals before they are eventually routed using Out (line 14). Mix sums the signals that are present in its argument array ("sample by sample") to create a monophonic signal. Note that, from a technical point of view, Mix is not a UGen and, even if it does have *ar and *kr methods, it is preferable to use the new method instead (the former methods are deprecated and will most likely be obsolete in some future release). Note also that the overall amplitude has to be scaled in order to avoid distortion.

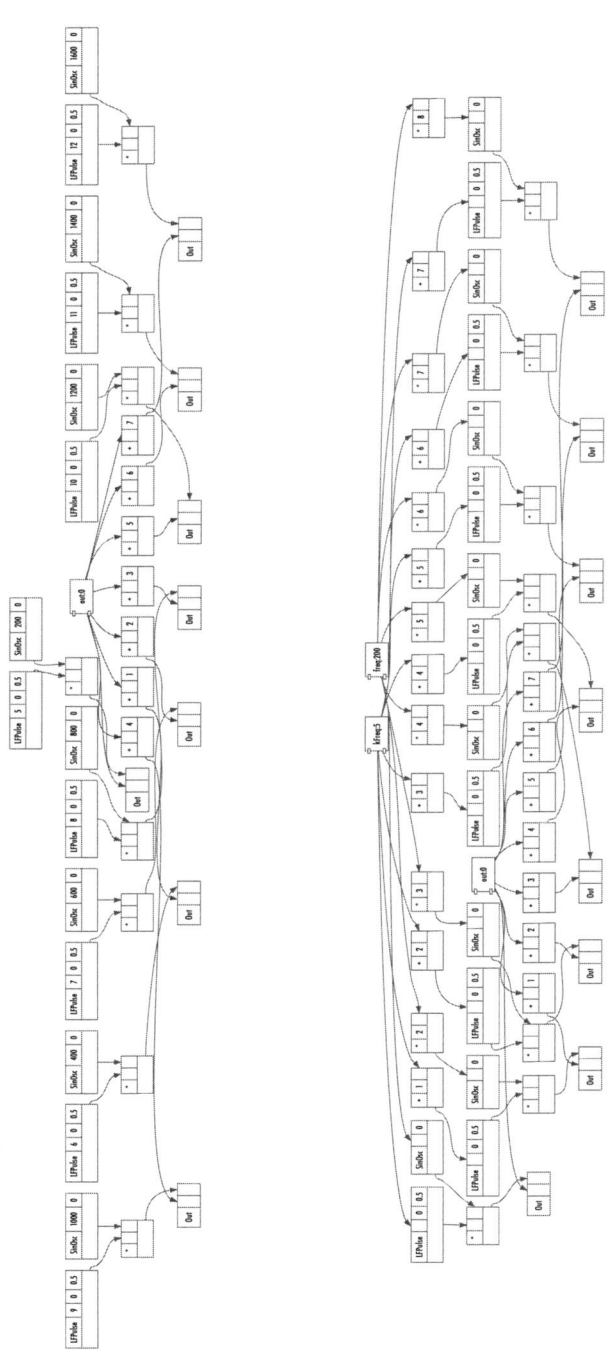

Figure 6.9   Two UGen graphs in a case of cycle.

```
1   SynthDef(\chan , { arg out = 0, freq = 200, kFreq = 5 ;
2       var n = 8;
3       var arr = Array.newClear(n);
4       n.do({|i|
5           arr[i] = LFPulse.ar(i+kFreq)*SinOsc.ar(i+1*freq)
6       }) ;

8       Out.ar(out, Mix(arr)/n) ;
9   }).add ;

11  s.scope ; // 1 chan
12  Synth(\chan ) ;
```

Using the fill method of Array, we can further shrink the code like this:

```
1   SynthDef(\chan , { arg out = 0, freq = 200, kFreq = 5 ;
2       var n = 8;
3       var arr = Array.fill(n, {|i| LFPulse.ar(i+kFreq)*SinOsc.ar(i+1*freq)
4       }) ;
5       Out.ar(out, Mix(arr)/n) ;
6   }).add ;
```

In fact, we can shrink it even more, by turning to Mix's fill method.

```
1   SynthDef(\chan , { arg out = 0, freq = 200, kFreq = 5 ;
2       var n = 8;
3       var mix = Mix.fill(n, {|i| LFPulse.ar(i+kFreq)*SinOsc.ar(i+1*freq)}) ;
4       Out.ar(out, mix/n) ;
5   }).add ;
```

The UGen graph of the previous SynthDef is partially depicted in Figure
6.10[3].

---

[3] It is apparent that Mix is not a UGen: it is not present in the graph. Internally,

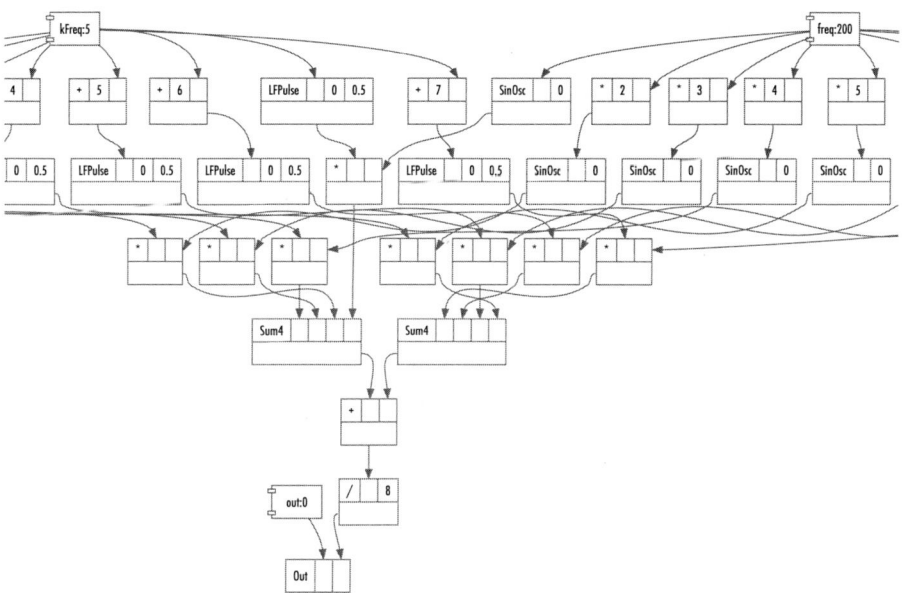

Figure 6.10   Graph of the SynthDef "chan".

6.7 Multichannel Expansion

We will conclude this chapter with a discussion on the "multichannel expansion" feature. Consider the following example:

---

a series of Sum4 UGens are used, which are mixers optimized for blocks of 4 signals.

```
1  SynthDef(\pan , {arg out;
2      Out.ar(out,
3          Pan2.ar(SinOsc.ar, MouseX.kr(-1,1))
4      )
5  }).add ;

7  s.scope ;
8  Synth(\pan ) ;
```

As shown in the stethoscope window, the mouse's x dimension controls
the stereophonic positioning of the sinusoidal signal. The arguments of the
Pan2 UGen —which is responsible for stereophonic panning— are in, position and level, which correspond to: the signal to be positioned in the stereo
image; its spatial positioning encoded as a float number in the $[-1, 1]$ range
(-1 maps to left and 1 maps to right); an amplitude scale factor. The UGen
Pan2 would then correctly position a signal in the stereo field. That is to say
that Pan2 appropriately distributes a monophonic signal on two consecutive
outputs. Such a stereo signal is nothing more but a pair of signals to be indi-
vidually processed and, eventually, listened to in parallel —typically using two
speakers. It has been already discussed that conventional stereophonic configurations map public busses 0 and 1 to the left and right channels, respectively.
Yet, in the above SynthDef only the index of the first public bus (channel 0)
is provided as an argument to Out. The latter, then, automatically distributes
the two input signals to both 0 and 1 busses. This feature of SuperCollider
is referred to as *multichannel expansion*. Figure **6.11** shows the resulting UGen
Graph. As shown, the Out UGen receives as a second argument (numChannels)
an array of signals —see also its help file. Hitherto, we have used the Out UGen
with monophonic signals alone, yet, it can be used with arrays as well —as in
the case of Pan2's output. Consider the graph in Figure **6.11**. There can be no
doubts: Pan2's output is an array of two distinct signals. Consider also the next
code snippet, where we use the typical array notation to access the individual
signals.

```
1  s.scope ;
2  {SinOsc.ar([60.midicps,  69.midicps])}.play ;
3  {SinOsc.ar([60.midicps,  69.midicps][0])}.play ;
4  {SinOsc.ar([60.midicps,  69.midicps][1])}.play ;
```

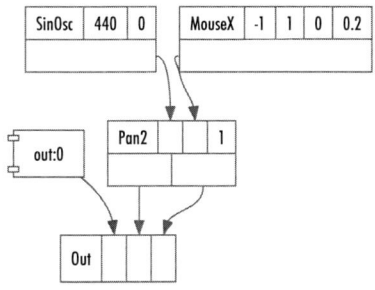

Figure 6.11   Graph of the SynthDef "pan2".

In the following example there are three versions of the same SynthDef:

```
1  // 1. explicit routing
2  SynthDef( "multi1", {
3      Out.ar(0, SinOsc.ar(300)) ;
4      Out.ar(1, SinOsc.ar(650)) ;
5  }).add ;

7  // 2. channelsArray in Out
8  SynthDef( "multi2", {
9      Out.ar(0,
10         [SinOsc.ar(300), SinOsc.ar(650)]
11     )
12 }).add ;

14 // 3. array in an argument
15 SynthDef( "multi1", {
16     Out.ar(0,  SinOsc.ar([300, 650]))
17 }).add ;
```

In the first case, two signals are explicitly written on bus 0 and 1. In the second case, Out's channelsArray argument is passed an array of two signals which are written to channels 0 and 1 respectively. The third case is technically identical with the second, only that it is described using a more expressive language style. Here, an array of frequencies is passed as an argument to a singleton SinOsc UGen. Generally speaking, every time an array is passed as an argument to some UGen, SuperCollider will implicitly expand the outer UGen so that a new instance for each of the corresponding element of the array is created. The output of that UGen will, in turn, be a new array. Note, however, that multichannel expansion occurs exclusively with instances of Array and not in the case of its' super/sub-classes. Therefore, the three cases above are technically identical, in that they result to the very same UGen graph, as shown in Figure **6.12**.

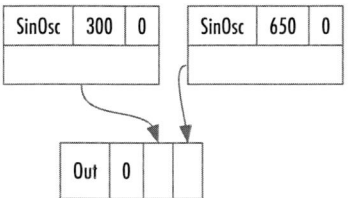

Figure 6.12   Graph of the SynthDefs with multichannel expansion.

The following example demonstrates how multichannel expansion may be exploited to control complex processes.

```
1  (
2  SynthDef(\sum , { arg out = 2 ;
3      Out.ar(out, Array.fill(16, {
4          SinOsc.ar(
5              freq:Rand(48, 84).midicps,
6              mul: LFNoise1.kr(Rand(2, 0.1))
7          )
8      })
9      )
10 }).add ;


13 SynthDef(\mix , {arg in;
14     Out.ar(0, Mix(In.ar(in, 16)/16))
15 }).add ;
16 )

18 b = Bus.audio(s, 16) ; b.scope ;
19 x = Synth(\sum , [\out , b]) ;
20 Synth.after(x, \mix , [\in , b]) ;
```

The output of the "sum" SynthDef is an array of 16 channels. If appropriate equipment is available, the latter would represent actual audio signals routed to speakers. Each signal is a sinusoidal wave with a random frequency selected from a 3 octaves tempered scale (starting from the bass-clef C, or 48 in midi notation). The amplitude of each channel is controlled by an LFNoise1 object that is constrained in the $[0, 1]$ output range and which is operating at some arbitrary frequency between 2 and 0.1 Hz (so, a frequency of 0.1 indicates a period of 10 seconds). The frequency is set using a Rand UGen which generates a single random number on every instantiation of a new Synth. The "mix" SynthDef is an auxiliary mixer for a 16-channel audio bus —note that a 16-channel audio bus is expected in the In's input. Accordingly, such a bus is registered and subsequently visualized in line 18 (note how scope can be directly invoked for instances of Bus). Synth x, then, writes 16 channels to the multichannel bus b (as illustrated in the GUI). The mixer then reads from that bus, sums up all channels and eventually writes a monophonic signal to the soundcard's first output.

Consider the following example, where we are explicitly generating an array of 16 signals.

```
1  SynthDef("multi16mixPan", { arg bFreq = 100 ; // base freq
2               var left, right ;
3               var sigArr = Array.fill(16,
4                   { arg ind ;
5                   var index = ind+1 ;
6                       Pan2.ar(
7                           in:    SinOsc.ar(
8                               freq: bFreq*index+(LFNoise1.kr(
9                                   freq: index,
10                                  mul: 0.5,
11                                  add: 0.5)*bFreq*index*0.02) ,
12                              mul: 1/16 ;
13                              ),
14                          pos: LFNoise1.kr(freq: index)
15                          )
16                  }) ;
17           sigArr.postln ;
18           sigArr = sigArr.flop.postln ;
19           left = Mix.new(sigArr[0]) ;
20           right = Mix.new(sigArr[1]) ;
21           Out.ar(0, [left, right])
22           }
23 ).add ;

25 a = Synth(\multi16mixPan ) ;

27 c = Bus.control ;
28 a.map(\bFreq , c) ;

30 x = {Out.kr(c, LFPulse.kr(
31     MouseX.kr(1,20), mul:MouseY.kr(1, 100), add:250).poll)}.play ;
32 x.free ;
33 x = {Out.kr(c, Saw.kr(MouseX.kr(0.1, 20, 1), 50, 50))}.play ;
```

The most complicated part of the code is between lines 7–13, where control UGens are associated with oscillators —it is left as an exercise to the reader to analyze this part of the code. Remember that sigArr includes 16 stereo signals, each of them being the output of a Pan2. Then, for each sinusoid, pos varies randomly between $[-1, 1]$ at a frequency equal to index (line 14). The bigger the index , the higher the frequency of the oscillator and, therefore, the faster the modulation of the signal's spatial positioning. Note that LFNoise1 UGen is used, which interpolates between successive values, so that no sudden "jumps" occur in the stereo image.

Upon evaluation, a message similar to the one below is expected in the post window.

```
1    [ [ an OutputProxy, an OutputProxy ], [ an OutputProxy, an OutputProxy ],
2        [ an OutputProxy, an OutputProxy ], [ an OutputProxy, an OutputProxy ],
3        [ an OutputProxy, an OutputProxy ], [ an OutputProxy, an OutputProxy ],
4        [ an OutputProxy, an OutputProxy ], [ an OutputProxy, an OutputProxy ],
5        [ an OutputProxy, an OutputProxy ], [ an OutputProxy, an OutputProxy ],
6        ...etc...

8    [ [ an OutputProxy, an OutputProxy, an OutputProxy, an OutputProxy,
9            an OutputProxy, an OutputProxy, an OutputProxy, an OutputProxy,
10           an OutputProxy, an OutputProxy, an OutputProxy, an OutputProxy,
11           an OutputProxy, an OutputProxy, an OutputProxy, an OutputProxy ],
12       [ an OutputProxy, ...etc...
```

SuperCollider refers to the output of some UGen with multiple outputs (such as Pan2) as an OutputProxy. Albeit not relevant here, it is interesting to note that each element of the resulting array is in turn another instance of Array composed of two elements, one per audio channel. The first postln is invoked in line 17 and represents an array of 16 arrays, each containing 2 signals. The method flop, then, rearranges the array so that it now comprises of 2 arrays containing 16 signals each —see the second printed block. The arrangement of the elements in sigArr is therefore

```
[[sig0sx, sig0dx], [sig1sx, sig1dx], [sig3sx, sig3dx]
        ..., [sig15sx, sig15sx]]
```

Invoking the flop method will result in a new array with the following structure:

```
[[sig0sx, sig1sx, sig2, sx, sig3sx, ..., sig15sx],
 [sig0dx, sig1dx, sig2, dx, sig3dx, ..., sig15dx]]
```

It is composed of 2 arrays —one per channel— each of which contains an array made up of the first or second element of each of the original 16-element arrays. It is then possible to mix the two arrays in two individual monophonic signals which are assigned to the variables left and right (lines 19–20), respectively. Then, passing the [left, right] array to Out results again in a multichannel expansion so that the first signal will be sent to the left speaker via bus 0 and the second to the right via bus 1. In the rest of the code, an instance of this synth is created and a bus is used to map the output of a MouseX to control bFreq. The reader should be able to understand this code easily by now.

Figure **6.13** shows the UGen graph of the "multi16mixPanGraph" SynthDef.

6.8 Conclusions

In this chapter, under the broader topic of "control", a series of different, albeit relevant, topics have been examined. By now several of the most important aspects of SuperCollider have been already dealt with. The next chapter is again dedicated to control structures, but this time the focus is on the temporal organization of events.

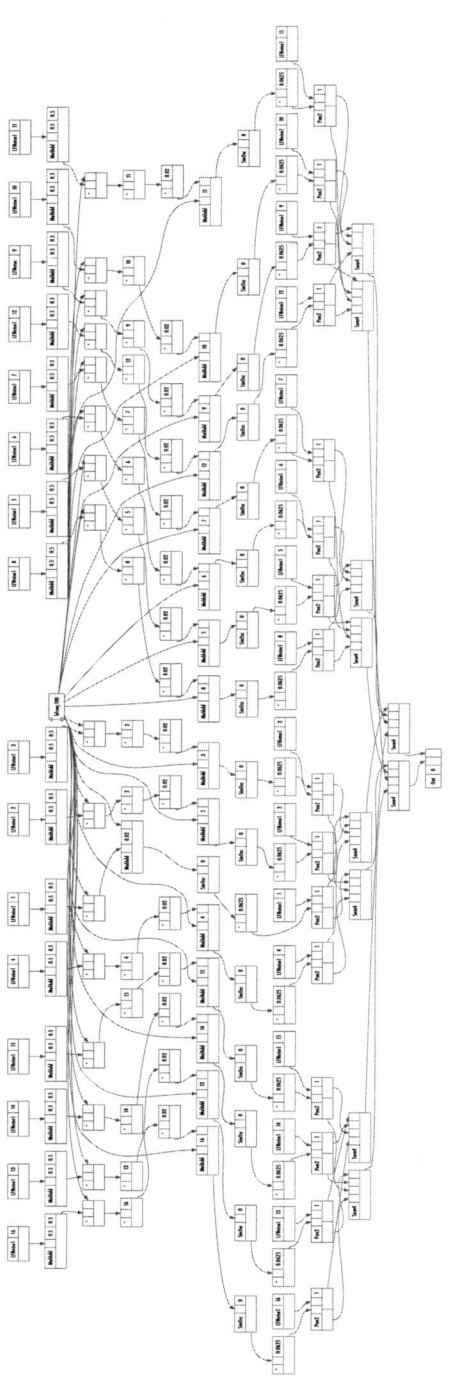

Figure 6.13 Graph of the SynthDef "multi16mixPan".

7 Organized sound: scheduling

Over the previous chapters, many aspects of SuperCollider have been introduced. Starting from UGens provided by SC and the possibility of building patches, it is already possible to experiment with sound generation and with interactive control in real-time. Before briefly introducing the implementation in SC of the most common techniques for sound synthesis, it is appropriate to discuss certain aspects of time management. SuperCollider excels in this task when compared to other programming environments because it seamlessly integrates audio synthesis and algorithmic composition. The control of information and its processes, through the SC language, and audio signal synthesis, through the management of UGens using SynthDefs, are two key ideas that have to be understood: borrowing an expression by Edgar Varèse, it is time to get to "organized sound". The term is interesting because it combines the work on sound matter and its temporal organization: it is a very general definition of both music and composition, intended as a set of sound events. In very general terms, 'scheduling' is precisely the allocation of resources for the realization of an event at a certain time[1]. As usual in SC, there are many different ways to manage scheduling.

7.1 Server-side, 1: through UGens

---

[1] For example, http://en.wikipedia.org/wiki/Scheduling

One option that comes from analog synthesizers is to manage the sequencing of events directly through audio signals. The following example, despite its simplicity, shows an aspect that has already been discussed: an envelope applied to a continuous signal can transform the signal into a set of discrete events.

```
1  (
2  SynthDef.new("pulseEvent", {|seqFreq = 4, sawFreq = 0.125, sinFreq = 0.14|
3      Out.ar(0,
4          Pulse.kr(seqFreq, width:0.1, mul:0.5, add:0.5)
5          *
6          Formant.ar(fundfreq:(
7              LFSaw.kr(sawFreq, mul:6, add:60)
8              +
9              SinOsc.ar(sinFreq, mul:3)
10         ).round.midicps),
11     )
12 }).add ;
13 )

15 x = Synth.new("pulseEvent", [\seqFreq , 4]) ;
16 x.set(\seqFreq , 5) ;
17 x.set(\sawFreq , 1/4) ;
18 x.set(\sinFreq , 1/2) ;
```

In this case, the envelope is provided by a signal in the form of a unipolar square wave ($[0, 1]$), with a duty cycle of 10% (0.1). The signal "windows" a signal generated by the UGen Formant, which in turn generates a complex harmonic signal from a fundamental frequency, plus another set of frequencies to be specified (on the model of the human voice). The frequency of the latter is given by a complex control curve that results from the interaction of a sawtooth wave (an iterating ramp) and by a sine wave (that increases/decreases it). Through round the curve, which is meant to represent pitches expressed in MIDI notation, is "jagged" and converted into frequency. Note that the sequence of pitches depends on the interrelation among the three frequencies seqFreq, sawFreq, sinFreq, as shown by lines 16-18.

The next example (which the reader is invited to implement in another way) exploits the UGen InRange. The latter returns a signal whose value depends on the inclusion (or not) of the value of the input sample in the specified range (here $[0, 1]$). If the value is inside the range, InRange returns the same value, otherwise it returns 0. The method sign returns 1 if the signal is positive, 0 if it is 0. Therefore, the output signal will consist of 1 or 0 if the input exceeds or not the threshold of 0.35. A similar envelope signal creates "holes" in the audio signal every time the value is 0, therefore transforming a continuous signal into a set of events.

```
1  (
2  SynthDef.new("pulseEventThresh", {
3      |seqFreq = 4,      sawFreq = 0.125,      sinFreq = 0.14|
4      Out.ar(0,
5          InRange.kr(LFNoise0.kr(15), 0, 1).sign
6          *
7          Formant.ar(fundfreq: (
8              LFSaw.kr(sawFreq, mul:6, add:60)
9              +
10             SinOsc.ar(sinFreq, mul:3)
11         ).round.midicps),
12     )
13 }).add ;
14 )

16 x = Synth.new("pulseEventThresh") ;
```

The following code shows what happens in 3 seconds.

```
1  {InRange.ar(LFNoise2.ar(15), 0, 1).sign}.plot(3)
```

If we evaluate the line 15 more times, it becomes clear that multiple synths are created. While obvious, this is a general point in relation to scheduling: parallelism (voices, layers, as the reader may want to refer to them) is handled by SC simply creating (literally 'expanding') more synths, the result of interactively evaluating the code. For example, the following code uses a variant of the previous SynthDef to instantiate 100 parallel synths:

```
 1  (
 2  SynthDef.new("schedEnv", {
 3      Out.ar(0,
 4      Pan2.ar(          // panner
 5          InRange.ar(LFNoise2.ar(10), 0.35, 1).sign
 6              *
 7          Formant.ar(LFNoise0.kr(1, mul: 60, add: 30).midicps, mul:0.15),
 8      LFNoise1.kr(3, mul:1) ) // random panning
 9      )
10  }).send(s) ;
11  )

13  (
14  100.do({
15      Synth.new("schedEnv") ;
16  })
17  )
```

As shown, one (1) linguistic expression (100.do) results in 100 (one hun-dred) synths. And they might be more, computational resources permitting. From an audio perspective, as the multiplying factor jumps from 0 to 1 and vice versa, there is a strong pulsing background noise (that remotely resembles a sort of old vinyl).

The next example –even if very simple– is an example of procedural sound design in SC. Usually, the sound designer for cinema and video (the gaming sce-nario requires more innovative perspectives) works in non-real-time, by sound editing in a DAW (Digital Audio Workstation). This type of work is strictly related to "fixed media", the output being a coded text/object (just think of a film). But other productive situations require a different type of organization, and impose other constraints: for example, an interactive installation asks for audio generation on demand, that not only has to be controllable in real-time, but maybe also without a predefined duration. The following code is dedicated to model a very typical sound, that even if certainly not very complex, played a pivotal role in the soundscape of modernity: the phone signal, and in par-ticular the Italian one. In the absence of an accesible technical specification of its features (even though there must be details somewhere), we may proceed empirically. Analyzing a recording of the sound, we observe that, first, the sig-nal is a sine wave with a frequency of 422 Hz (between A and A♭, you can check this by evaluating 68.midicps= 415.30469757995 and 69.midicps= 440.

The Italian signal representing 'busy' is the simplest sound imaginable: it is a regular sequence of pulses and silences, regularly spaced with 200 ms. The sequencing can easily be obtained with an unipolar square wave that acts as a "gate": $[0, 1]$. Since each half cycle must be equal to 200 ms, then the period will be $200 \times 2 = 400$ ms. The implementation simply uses the UGen LFPulse enveloping a SinOsc (7). The case for the dial tone is a bit more complex temporally. By analyzing the signal a more musical pattern emerges, one that has been shaped by means of an envelope, also taking into account that the basic time unit is 1. Empirically (by ear), the time segments do not look exactly proportional. This has prompted some adjustments (see the durations 2.5 and 1.5) (15): If we consider the factor timescale then the envelope takes 1.2 seconds (i.e. env.sum * 1/5). The repeat period is instead 2 seconds (cycleDur), which in turn determines the frequency of the trigger (1/cycleDur).

```
1   // Phone signal (Italian)

3   (
4   // busy: pulses at 200 msec intervals
5   {
6       var freq = 422, dur = 0.2;
7       LFPulse.kr(1/( dur*2))*SinOsc.ar(freq)
8   }.play ;
9   )

11  (
12  // free
13  {
14      var freq = 422, cycleDur = 2 ;
15      var env = Env([0,1,1,0,0,1,1,0, 0], [0,1,0, 1, 0, 2.5, 0, 1.5]);
16      EnvGen.kr(env, 1,
17          gate: Impulse.kr(1/cycleDur),
18          timeScale:1/5, doneAction:0)
19      * SinOsc.ar(freq)
20  }.play
21  )
```

Apart from the exercise of modeling per se, which helps to understand the organization of sound materials, a possible application could be applied to an interactive installation in which a certain condition determines if the phone is free or busy.

The UGen Select.kr (which, arrays) implements on the server side a typical function of sequencing: using control rate (kr), each time a new value is calculated a new element which is selected from the array array.

```
1  (
2  SynthDef(\select , {
3      var array ;
4      array = Array.fill(64, {|i| (i%4) + (i%7) + (i%11) + 60}) ;
5      array = array.add(90).postln.midicps;
6      Out.ar(0,
7          Saw.ar(
8              Select.kr(
9                  LFSaw.kr(1/6).linlin(-1.0,1.0, 0, array.size),
10                 array
11             ),
12             0.2
13     ));
14 }).add;
15 )

17 Synth(\select ) ;
```

The previous code has two relevant aspects. First, array is built using a variable (operator) modulo for the counter i. The result is the sequence of pitches in Figure **7.1**. The last element (90) has been added to provide a clear acoustic index for the end of the cycle.

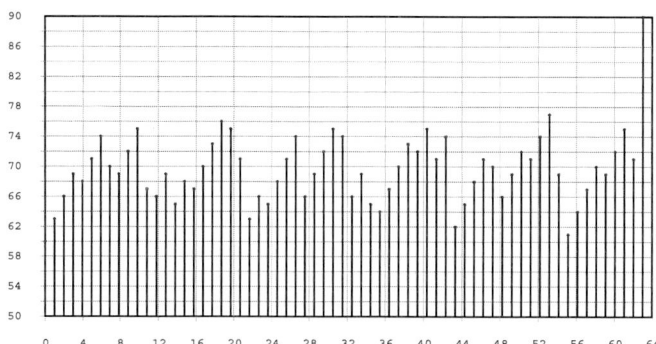

Figure 7.1   Sequence built with multiple applications of the modulo operator.

Select controls the frequency of Saw. In output, it cycles on array in relation to a signal providing it with indexes. In this case the signal is LFSaw, which is a ramp (a signal which increases linearly between a minimum and a maximum, and then is reset to zero): LFSaw is scaled by means of the operator linlin so as to cycle through all the indices of the array (0, array.size). Its periodicity indicates by how much the ramp is increasing, in this case, 6 seconds. The frequency is in fact $\frac{1}{6}$, expressed in order to make it immediately apparent that the period is $T = 6^2$. It is interesting to note that the use of ramp signals is typical in analog synths[3], where scheduling is often managed through appropriately generated, continuous signals.

7.2 Server side, 2: Demand UGen

A UGen such as Sequencer introduces in the generation of an audio signal (in a SynthDef) aspects of control typically associated to a higher level. In this direction, SC implements a very unique approach to scheduling by means of the so-called Demand UGens, where the term refers, as we will discuss, to "demand rate". The UGen Demand.ar(trig, reset, [..ugens..]) operates in relation to a trigger (trig). Whenever a trigger signal is received[4], the UGen requires a value from each of the other UGens in the array [..ugens ..]. These UGens must be a certain type, in this case Demand, as they generate a value (and only one) on request. In the following example, the trigger signal to Demand is generated by the UGen Impulse. Each of the generated pulses provides a transition between a minimum (0) and maximum (here 1), which act as a trigger. There is no point in discussing the synthesis process (line 7), except to note that the frequency of Pulse is handled by freq. Line 6 assigns a value to freq via the UGen Demand. Whenever a trigger is received, Demand asks for the next value

---

[2] Another UGen that allows to generate discrete events in a similar way is Stepper.

[3] The most common implementation is called a "phasor": see the UGen with the same name in SC and the next chapter.

[4] Remember that a triggering event occurs whenever a signal moves from 0 a positive value.

in the demand array a. The latter is filled with a Dseq UGen (line 4): it generates a sequence of values as specified in the list provided as its first argument ([1,3,2,7,8]), and repeated –in this case– 3 times. By listening to the result, it is clear that the sequence a is made up by the repeating 3 times a segment: when the segment is completed, Demand returns the last value in the sequence. By the way, at the synthesis level it is interesting to notice how the audio signal implements a kind of chorus effect. It is the sum of 10 square waves oscillating in their frequency with a random variation in the range $[0, 5]$ Hz: a kind of variable pitch (but not enough to introduce dissonance) typical of many instruments playing together.

```
1   (
2   {
3       var a, freq, trig;
4       a = Dseq([1, 3, 2, 7, 8]+60, 3);
5       trig = Impulse.kr(4);
6       freq = Demand.kr(trig, 0, a.midicps);
7       Mix.fill(10, { Pulse.ar(freq+5.rand)}) * 0.1

9   }.play;
10  )
```

Thus, Demand UGens are essentially generators of values on request, and differ from each other for the sequences of values that they can produce. These sequences are encoded in an interesting form: they are described in the form of a "pattern". In fact, the sequence is not described per se, but by defining a production rule. In the previous example the generated sequence would be:

```
1 [ 61, 63, 62, 67, 68 , 61, 63, 62, 67, 68, 61, 63, 62, 67, 68 ]
```

A pattern is therefore a form that is described by a procedure. For example, Dseq generates sequences of values by iterating n times an array, where n can be infinite, a value that SC represents through the reserved word inf. UGens as Drand describe a different pattern. Drand expects two arguments: the first is an array of values, the second (repeats) a number that represents the number of values picked up randomly from the provided array. In the example, the array is the same as in the previous case, and freq is a sequence with the length inf of values taken pseudo-causally from the same array a. In addition, the triggering frequency has been increased from 4 to 10.

```
1  (
2  {
3      var a, freq, trig;
4      a = Drand([1, 3, 2, 7, 8]+60, inf);
5      trig = Impulse.kr(10);
6      freq = Demand.kr(trig, 0, a.midicps);
7      Mix.fill(10, { Pulse.ar(freq+5.rand) }) * 0.1

9  }.play;
10 )
```

The result is a kind of improvisation on a finite set of pitches (a mode, one could say). The expressive power of Demand type UGen lies in recursive nesting. The following example is quite similar to the first case discussed here, except that one of the elements of the array on which it operates Dseq uses a Dxrand.

```
1  (
2  x = {|trigFr = 1|
3      var freq, trig, reset, seq;
4      trig = Impulse.kr(trigFr);
5      seq = Dseq(
6          [42, 45, 49, 50,
7          Dxrand([78, 81, 85, 86], LFNoise0.kr(4).unipolar*4)
8          ], inf).midicps;
9      freq = Demand.kr(trig, 0, seq);
10     Pulse.ar(freq + [0,0.7] + LFPulse.kr(trigFr, 0, 0.1,freq*2))* 0.5;
11 }.play;
12 )

14 x.set(\trigFr , 10) ;
```

In this case seq is a sequence that repeats infinite number of times (inf) a pattern consisting of the numbers $42, 45, 49, 50$, and a fifth item defined by

Dxrand: a cousin of Drand, this UGen randomly chooses a value from the provided array ([78, 81, 85, 86]), but ensures that the output sequence does not include adjacent repetitions of the same element. When Demand reaches the element of the array Dseq consisting of Dxrand, its encoded procedure is performed, generating a number of elements equal to repeats. The size of the output sequence (i.e. the value of repeats) is controlled by LFNoise0: in essence, it varies pseudo-randomly in the range [0, 4]. If the minimum is chosen, the contribution of Dxrand is zero, if the maximum is picked up, then the sequence consists of all four values, in pseudo-random order but with no adjacent repeats. From the point of view of audio, the presence of the array [0,0.7] in the argument freq results in multichannel expansion: the right channel will be "detuned" by 0.7 Hz. In addition, LFPulse produces a kind of grace note an octave above the initial tone, which is coupled with the triggering of the demanded event through trigFr.

```
1  (
2  SynthDef("randMelody",
3      { arg base = 40, trigFreq = 10;
4      var freq, trig, reset, seq;
5      var structure = base+[0, 2, 3, 5, 6, 8] ;
6      trig = Impulse.kr(trigFreq);
7      seq = Dseq(
8          structure.add(
9          Dxrand(structure+12, LFNoise0.kr(6).unipolar*6))
10         , inf).midicps;
11     freq = Demand.kr(trig, 0, seq);
12     Out.ar(0,
13         Mix.fill(5, {Saw.ar(freq +0.1.rand + [0,0.7])* 0.1}));
14 }).add;
15 )

17 x = Synth.new("randMelody") ;
18 x.free ;

20 (
21 15.do({ arg i ;
22     Synth.new("randMelody",
23         [\base , 20+(i*[3, 5, 7].choose), \trigFreq , 7+(i/10) ])
24 })
25 )
```

In the previous example a SynthDef (constructed in a similar manner to previous examples) provides as arguments base and `trigFreq`: the first is the base frequency in MIDI notation, the second the triggering frequency. Here `Dxrand` adds to the sequence of pitches the same sequence but shifted an octave up, with the already discussed mechanism of random selection of pitches and length (9). In relation to audio synthesis, it is a simple variation of the previous example(13), as can be heard by instantiating a synth (17). The next cycle overlaps 15 voices: in each voice, the base frequency is increased by i multiplied by a value chosen from [3, 5, 7] (musical intervals of: a minor third, fourth and fifth). In addition, at each iteration (for each synth) the triggering frequency (7) increases by an amount equal to i/10. In this way a progressive *dephasing* is obtained: timings of the layers are slightly different, and the initial synchronization gradually disappears.

The idea at the basis of Demand UGen is to provide the ability to include within a SynthDef aspects typically associated with a higher level of control. A SynthDef becomes a sequencer for all purposes. Moreover, the possibility of nesting UGens is extremely powerful. But it maybe argued that it is conceptually straightforward to separate two aspects, which typically work at different rates: synthesis (audio rate) and scheduling (event rate). Not surprisingly, Demand UGens are closely related to the so-called "Patterns", language-side, high level, data structures for composition. In fact, Demand UGens are a sort of server-side counterpart of Patterns, which will be discussed later.

7.3 Language-side: Clocks and routines

The most usual (and perhaps more conceptually straightforward) way to perform the scheduling of events is to manage it from the language side. In order to define the execution of events in time, SC provides the abstract class `Clock` from which three classes inherit, `SystemClock`, `TempoClock` and `AppClock`. Therefore, an event can be scheduled as follows:

```
1  (
2  "waiting for 3 seconds".postln ;
3  SystemClock.sched(3.0, { "done".postln });
4  )
```

The method sched available for the class SystemClock requires as two arguments, a time interval and a function that will be evaluated after it. By evaluating the code, we get:

```
1  waiting for 3 seconds
2  SystemClock
3  done
```

The interpreter immediately executes line 2 (and prints on the post window), then moves to the next line and evaluates it, completing the interpretation process: it can be seen that SystemClock is immediately printed on the screen, that is, the object that is returned by the last espression. The scheduling is started, after 3 seconds the function is evaluated, and "done" is printed on the post window. The three types of clock differ in priorities and versatility. SystemClock is the clock with the highest priority and implements a concept of absolute (not musical) time, like a chronograph. TempoClock has a working mode similar to SystemClock but additionally it enables the user to manage a concept of musical time/tempo. TempoClock assumes that the default time-unit of 1 beat lasts 1 seconds (tempo = 60 bpm). In this way, time as calculated with reference to tempo coincides with absolute (chronographic) time. But the tempo property of TempoClock can be changed at will, according to a desired music tempo. In the following example, three clocks are in action, a System-Clock and two instances of TempoClock with different tempos. Note that tempo is not specified in beats per minute (bpm), rather as beats per second: So, 240 bpm equivalent to 240/60 "bps". Lines 7-9 manage the scheduling, and by evaluating them it becomes evident that the time value of 4.0 depends on, is relative to, the defined tempo.

```
1  (
2  "go".postln ;
3  // two TempoClocks: t, u
4  t = TempoClock(240/60) ;
5  u = TempoClock(120/60) ;
6  // scheduling
7  d = thisThread.seconds;
8  SystemClock.sched(4.0, { "done with chronometer".postln; });
9  t.sched(4.0, { "done with bmp 240".postln });
10 u.sched(4.0, { "done with bpm 120".postln });
11 )
```

Finally, AppClock is a clock with low priority to be used for scheduling GUI-related events. In SC, GUIs (must) have a lower priority than audio. This means that the calculation of audio has always priority over GUIs, which are updated only if computational resources are available. GUI elements should therefore be scheduled via AppClock.

The use of the subclasses of Clock is at the basis of scheduling in SC, but typically not directly: rather, other data structures are available to control scheduling in relation to a clock. The fundamental control structure for scheduling in SC is the "routine". By themselves, routines are data structures that extend the modus operandi of functions, and their use is not limited to scheduling: that is, scheduling is only one of the possible applications of routines, although they are often used in this role. The following example shows a minimal routine. As shown, the routine receives as an argument a function. The only unknown message in the code is wait, received by 2, which is a floating-point number. Intuitively, wait concerns time management, i.e. scheduling. The routine r defines a program to be executed. In the program, a message wait received by a number indicates a waiting time (equal to the receiver of the message) that pauses the execution before the next expression. There are three expressions (4-6): two are print requests, the third (between them) is the call to wait.

```
1  (
2  // Minimal routine
3  r = Routine.new({
4      "waiting 2 seconds".postln ;
5      2.wait;
6      "done".postln ;
7  }) ;
8  )

10 SystemClock.play(r) ;
```

The management of the function over time is controlled by a clock, here SystemClock: the clock is asked to schedule (play) the routine (r, 10). The clock interprets the objects receiving the message wait as an amount of time in which to suspend the execution of the sequence of expressions contained in the function body. When the execution is resumed, the expressions are evaluated as quickly as possible (i.e., at a rate determined by CPU cycles, which should be immediately).

One could argue that the example is not qualitatively very different from the previous one in which the method sched of SystemClock was in use. But there is a radical difference: by means of a routine, it becomes possible to implement a complex sequence of expressions, dynamically related to time. In the next example, the function contains a loop that for 10 times prints out "slowing down", each time assigning a time value of i * 0.1, that is, it determines the scheduling procedurally by referring to the counter i. The routine r waits for an amount time that is progressively increasing. Once the cycle is endend, r waits for 1 second, writes "the", waits for another second, and writes "end".

```
1  (
2  // Some more expressions
3  r = Routine.new(
4    {      var time ;
5      10.do ({ arg i ;
6          "slowing down".postln ;
7          time = (i*0.1).postln ;
8          time.wait ;
9      }) ;
10     1.wait ;
11     "the".postln ;
12     1.wait ;
13     "end".postln ;
14   }
15  ) ;
16  )

18  SystemClock.play(r) ;
19  r.reset ;
20  SystemClock.play(r) ;
```

Note that the routine "remembers" its internal state: if SystemClock.play(r) (18) is evaluated again, the post window shows SystemClock. The clock is returned because the routine has been completed. To reset the routine's internal state to its initial condition is necessary to send r the message reset (19). At this point, the routine can be run again (20).

As usual, the SC language allows us to write even more concise expressions, by relying on the so-called "polymorphism", i.e. the fact that certain methods have the same name but different semantics on different objects. Let us consider the following example.

```
1  (
2  // polymorphism, 1
3  r = Routine({
4      10.do{|i| i.postln; 1.wait}
5  })
6  )
7  r.play ;

9  // polymorphism, 2
10 { 10.do{|i| i.postln; 1.wait} }.fork ;
```

Upon defining the routine r (3-5), it can be executed by directly calling the method play on the routine itself. The method play is responsible for instantiating a clock (the default one, that is, TempoClock) that in turn will schedule the process. Yet, the method fork is defined directly on a function. The function (10) is then wrapped into a routine and the latter is executed on the default clock (with respect to functions, the user may be reminded of the "behind the scenes" processes activated by the interpreter in the case of the method play). As a side remark, it is useful to mention an interactive control element: a routine may be stopped by selecting stop from the menu *Language* (i.e. also with the usual key combination available to stop audio synthesis).

From these simple examples it should be clear that routines constitute the basis for the management of temporal processes, that can be of any complexity. Put simply, when the user wants to execute expressions according to a certain temporal organisation, it is enought to envelope them in a routine and provide the opportune wait expressions. In its simplest form, such a sequence of expressions can be thought of as a list to be performed in sequence, and exactly the same process as in a MIDI file or in a textual Csound "score" file. But obviously the most interesting aspect lies in the use of control structures in the function, together with the possibility of interacting with running processes by means of variables.

7.4 Clocks

The next example discusses generating a GUI that works as a simple stopwatch, and allows to introduce some more elements in the discussion about scheduling. By running the code, a small window pops up that displays the progress of time starting from 0: The clock starts and is stopped respectively when the window opens and closes.

```
1  (
2  var w, x = 10, y = 120, title = "Tempus fugit"  ; // GUI var
3  var clockField ;
4  var r, startTime = thisThread.seconds ; // scheduling

6  w = Window.new(title, Rect(x, y, 200, 60)) ;
7  clockField = StaticText.new(w, Rect(5,5, 190, 30))
8      .align_(\center )
9      .stringColor_(Color(1.0, 0.0, 0.0))
10     .background_(Color(0,0,0))
11     .font_(Font(Font.defaultMonoFace, 24));
12 r = Routine.new({
13     loop({
14         clockField.string_((thisThread.seconds-startTime)
15             .asInteger.asTimeString) ;
16         1.wait }) // the clock is updated each sec
17     }).play(AppClock) ;
18 w.front ;
19 w.onClose_({ r.stop }) ;
20 )
```

The first few lines declare the variables to be used. Basically the code provides two types of elements: GUI elements (lines 2, 3) and elements that handle time updating. In particular, line 4 assigns startTime the value of thisThread.seconds: thisThread is a peculiar object, a pseudo-environmental variable that contains an instance of the class Thread, which keeps track of how much time has passed since the beginning of the session of the interpreter. If you evaluate the expression thisThread.seconds a couple of times you will notice how the returned values increase accordingly (matching the seconds passed since the interpreter has started). Therefore, startTime contains a value that represents the time in which the code is evaluated (the moment 0 for the stopwatch). There is no point in dwelling in detail on the construction of the graphic elements, as it does not have anything special. Note only the style of programming in the

GUI that makes use of message chaining (lines 8-11) to set the different graphical properties. Note also that the font is defined by a specific object Font and
takes as the font name the string returned by Font.defaultMonoFace, that is, the
default monospaced font ("mono") in the operating system. Lines 12-17 define
instead a routine that calculates the new value to be displayed by the stopwatch
and updates the relative GUI element every second. The function in the routine contains an infinite loop. The infinite loop is defined by the value loop, that
is synonymous with inf.do. The cycle executes two expressions: the first one
updates the time display by setting the string of clockField (lines 13-14); the
second one requires the process to wait for a second before repeating the cycle
(1.wait, line 16). The new value for the time display is computed in three steps.
First the interpreter is asked to return the amount of time since it has started,
thanks to thisThread.seconds: this value is subtracted from the starting time
at initizalization phase, stored in startTime to retrieve the amount of time that
has passed by. The result is converted into integers (as seconds are the intended
resolution): therefore, we now have the number of seconds since the stopwatch
existed. Finally, the method asTimeString, defined for the class SimpleNumber,
returns a string in the form "hours: minutes: seconds". For example:

```
1  20345.asTimeString
2  05:39:05:000
```

The method play receives as an argument an object of type Clock but it
is not SystemClock: in fact, the latter can not be used in the event of a GUI.
AppClock must be used in its place. Finally, line 19 defines the property of the
window w: onClose takes as its argument a function that is executed when w is
closed. Here the function contains r.stop: the routine r is stopped when the
window closes.

The following code is a variant of the definition of routine r, demonstrat-
ing the last considerations. As we said, the use of AppClock is required because of computing priorities: in the case of scheduling, audio messages to the
server have priority over other features, including GUI. This would not allow
us to manage from the same routine the control of a synth and a GUI element.
However, such an operation is possible using the method defer available for
functions. In essence, the routine expressions concerning the GUI are collected

within a function to which the message defer is sent: in this way, their evaluation is deferred when computational resources are available (i.e. without affecting the computation of audio). In the example, GUI updating is enclosed in a function that is sent a defer message (Line 3-6): it is then possible to use SystemClock (8) to schedule the routine.

```
1  r = Routine.new({
2      loop({
3          {
4              clockField.string_((thisThread.seconds-startTime)
5                  .asInteger.asTimeString.postln) ;
6          }.defer ; // GUI elements must be "deferred"
7          1.wait })
8  }).play(SystemClock) ;
```

7.5 Synthesizers vs. events

Quite evidently, the previous approach can be extended to audio. Except for the problem of priorities in the management of the scheduling discussed earlier there is nothing specifically related to graphical interfaces. The following SynthDef generates a signal made up from the sum of 5 sawtooth waves whose frequencies are related harmonically. In addition, there is a "detune" component controlled by a sine wave which frequency depends on each harmonic component (through the counter i). The signal is enveloped by a ramp (6) that increases from 0 to 1 with a ratio of $\frac{9}{10}$ of its development, and decreases down to 0 for the last $\frac{1}{10}$. An important point to notice concerns the trigger: in this case, it is not generated by a UGen, rather it is passed from outside, by means of the argument t_trig. Once the trigger has value > 0, in order to shoot again, it should be reset (by sending another message) to a value of ≤ 0, which is quite inconvenient. But if the name of the argument indicating a trigger begins with t_, then by convention it is enough to send a single message to activate the trigger again, without forcing the reset.

```
1  (
2  SynthDef(\saw , { arg freq = 440, detune = 0.1, dur = 0.1, t_trig = 1;
3      Out.ar(0,
4          Pan2.ar(
5              Mix.fill(5, {|i|
6                  EnvGen.kr(Env([0,1,0], [dur*0.9,dur*0.1]), t_trig)
7                  *
8              Saw.ar(
9                  freq:freq*(i+1)+
10                 (SinOsc.kr((i+1)*0.2, 0, detune, detune*2)),
11                 mul:1/5)})
12     ))
13 }).add
14 )

16 (
17 x = Synth(\saw ) ;
18 ~mul = 1 ;
19 ~base = 60 ;
20 Routine.new({
21     var dur ;
22     inf.do {|i|
23         dur = [1,2,3,4,2,2].choose*~mul ;
24         x.set(
25             \t _trig, 1,
26             \freq , ([0,2,3,4,5,7, 9,10,12].choose+~base).midicps,
27             \detune , rrand(1.0, 3.0),
28             \dur , dur*0.95
29         ) ;
30         dur.wait ;
31     };
32     }
33 ).play(SystemClock)
34 )
35 // controlling interactively the routine
36 ~mul = 1/16; ~base = 72 ;
37 ~mul = 1/32; ~base = 48 ;
```

Once the synth x is built, the scheduling is run through an infinite routine (inf.do, 20). At each iteration, the routine sets (x.set, 22) the values for the synth's arguments. The routine regularly exploits pseudo-random values: for example, it chooses among a set of durations (21) and pitches (24) and it sets the detune in the range $[1.0, 3.0]$ (25). Two environment variables are used in the routine to handle the speed of events and the base pitch, respectively ~mul and ~base. In this way, it becomes possible to manage the scheduling interactively (e.g. by evaluating lines 34 and 35).

In the above example the basic idea is to build a synthesizer (a tool) and control it through a routine. The following SynthDef allows us to introduce a second, different approach. It provides a simple sine wave to which some vibrato and an amplitude envelope are added. The parameters of both can be controlled from the outside: a, b, c represent the envelope points, vibrato and vibratoFreq, two parameters for the vibrato.

```
1  (
2  SynthDef("sineMe1",{ arg out = 0, freq = 440, dur = 1.0, mul = 0.5, pan = 0,
3      a, b, c,
4      vibrato, vibratoFreq;

6      var env;
7      env = Env.new([0,a,b,c,0], [dur*0.05,dur*0.3,dur*0.15,dur*0.5], 'welch');
8      Out.ar(out,
9          Pan2.ar(
10             SinOsc.ar(
11                 freq: freq+SinOsc.kr(mul:vibrato, freq: vibratoFreq),
12                 mul:mul
13             ) * EnvGen.kr(env, doneAction:2)
14         ), pan)
15 }).add;
16 )
```

The amplitude envelope is used by a UGen EnvGen, its argument doneAction receives the value = 2. This means that, once the envelope is finished, the synth is deallocated from the server. This implies that the synth no longer exists: thus, the synth acts not so much as an instrument, rather as an event. Let us observe what happens in the routine:

```
1  (
2  var r = Routine.new({
3      inf.do({ arg i ;
4          var env, dur = 0.5, freq, end, mul, pan, vibrato, vibratoFreq ;
5          a = 1.0.rand ;
6          b = 0.7.rand ;
7          c = 0.5.rand ;
8          pan = 2.0.rand-1 ;
9          // 13 pitches on a non-octavizing modal fragment
10         freq = ([0,2,3,5,6,8,10,12,13,15,16,18].choose+70).midicps ;
11         dur = rrand(0.015, 0.5) ;
12         mul = rrand(0.05, 0.8) ;
13         vibrato = (dur-0.015)*100 ;
14         vibratoFreq = dur*10 ;
15         Synth.new(\sineMel , [
16             \vibrato , vibrato,
17             \vibratoFreq , vibratoFreq,
18             \a , a,
19             \b , b,
20             \c , c,
21             \freq , freq,
22             \dur , dur,
23             \mul , mul,
24             \env , env]
25             ) ;
26         end = 0.15.rand;
27         (dur+end).wait;
28     });
29 });
30
31 SystemClock.play(r);
32 )
```

At each iteration in the cycle, a synthesizer is created: it is deallocated when the envelope is concluded. In this second example, the object synth is not treated as an instrument, that is, as a persistent device (a trumpet, a bass, a violin) to be controlled in time. Rather, here the synth properly becomes a sound event, the equivalent of a musical note. The method doneAction: 2 shows (in this case) how it is most typically used. So, this is an example of a simple generative procedure that can last for an indefinite time, e.g. in the context of a permanent installation. Again, the SynthDef makes extensive use of pseudo-random values, that are responsible for the typical "whistled" quality of the sine wave.

The only important aspect is frequency control (line 10). The array defines a sequence of pitches resulting in a non-octaving mode of 13 pitches, stochastically selecting one of them, and then adding to it a value of 70 (base pitch).

The next example again takes into account the issue of signals from the technological soundscape, as we did before with the phone. The exact time signal emitted by the Italian National Broadcast Corporation, Radio RAI, is a signal of "synchronization and dissemination" which is called "Segnale orario RAI Codificato (SRC)" ("Hourly coded RAI signal") and generated by the National Institute of Metrological Research (INRIM). As indicated by the Institute:

> "The hourly time signal generated by the Institute and diffused by RAI (SRC) consists of a date code divided into two information segments, generated in correspondence with the seconds 52 and 53, and six acoustic pulses synchronized with the seconds $54, 55, 56, 57, 58$, and 00. The six acoustic pulses are each formed by 100 sinusoidal cycles of a 1000 Hz note. The duration of each pulse is 100 milliseconds[5]."

Figure **7.2** is the official encoding scheme for SRC provided by INRIM. The signal is formed by two blocks of binary data encoded with segments of 30 ms duration, in which a 2000 Hz sine wave represents 0, and a 2500 Hz sine wave represents 1. At 52 seconds, a first block of 32 bits (total: $32 \times 30 = 960$ ms) is emitted. It is followed at 53 seconds by a second block of 16 bits (total: $16 \times 30 = 480$ ms). Then 5 sinusoids are emitted, each of 100 ms (seconds $54 - 58$); second 59 is silent; finally, there is a last emission at second 60. As discussed, the 48 bits represent a set of time information, and the irregular alternation between the two frequencies of 2 and 2.5 kHz is responsible for the typical opening sound chirp.

The following example reproduces the SRC signal in SuperCollider. Lines 5 and 6 generate two random sequences of 32 and 16 bit. It would obviously be possible to generate the correct sequences, and in this way the subsequent implementation would not change. The two cycles (9-13 and 16-20) generate the two blocks of 960 and 480 ms. Here we chose to use synths as events. The generation of a sequence of 1000 Hz pulses follows (22). To differentiate the approaches, the first block of 5 pulses is obtained by enveloping a sine wave with a signal of type LFPulse, suitably parameterized. The last impulse is instead a synth-event of 100 ms. In all cases, the UGen Line is used essentially as a timed "deallocator" for the synth[6].

---

[5] http://www.inrim.it/res/tf/src_i.shtml

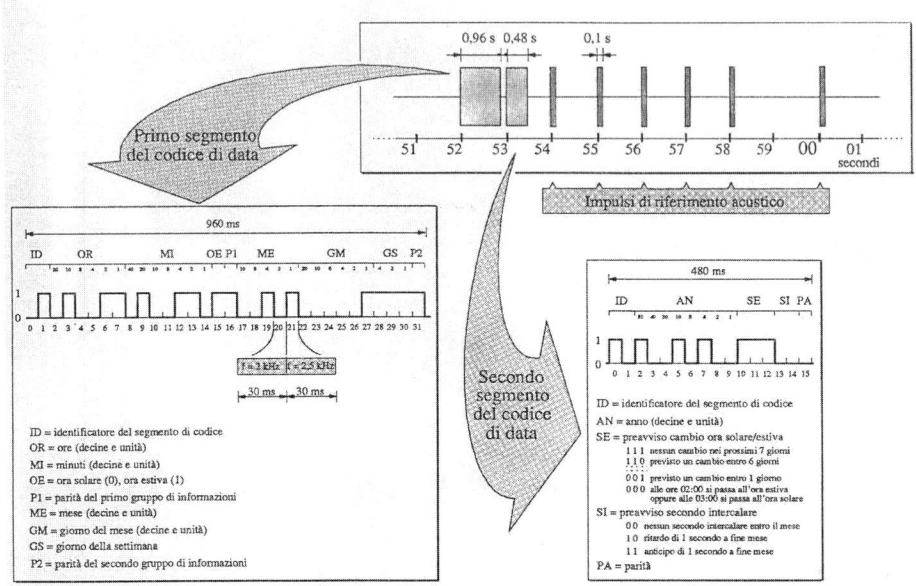

Figure 7.2 "Time diagram for the emissions of the various components of the SRC".

---

[6] From an acoustic perspective, both this signal and the previous phone example sound a bit too clean and bright. This is because they are usually heard through an analog transmission, frequently a low quality one. To get such an effect, it is possible e.g. to apply a filter or to add some background noise, as we will see in the next chapter.

```
1   (
2   // Hour signal SRC
3   {
4       var amp = 0.125 ;
5       var firstBit = Array.fill(32, {[0,1].choose}) ;
6       var secondBit = Array.fill(16, {[0,1].choose}) ;
7       var freq ;
8       // first 32 bits
9       firstBit.do{|i|
10          freq = [2000, 2500][i] ;
11          {SinOsc.ar(freq, mul:amp)*Line.kr(1,1,0.30, doneAction:2)}.play ;
12          0.03.wait
13      } ;
14      0.04.wait ;
15      // second 16 bits
16      secondBit.do{|i|
17          freq = [2000, 2500][i] ;
18          {SinOsc.ar(freq, mul:amp)*Line.kr(1,1,0.30, doneAction:2)}.play ;
19          0.03.wait
20      } ;
21      0.52.wait;
22      // 5 impulses with 1000 Hz
23      {
24          SinOsc.ar(1000, mul:amp)
25          *
26          LFPulse.ar(1, width:0.1)*Line.kr(1,1,5, doneAction:2)
27      }.play ;
28      6.wait;
29      // last
30      {SinOsc.ar(1000, mul:amp)*Line.ar(1, 1, 0.1, doneAction:2)}.play ;
31  }.fork ;
32  )
```

In the example, there is no explicit mention of the routine. In fact, the method fork defined for functions (that are the real conceptual and operational keys in SC) assumes that the function on which it is called is a scheduling one (that is, a function as those to be provided to routines). Therefore, fork incorporates the function into a routine and executes it with TempoClock.

Routines (and processes in general) can always take place in parallel. In the next example, 20 routines are running simultaneously. Pitches are generated by an array (pitches) which increases with a step of 4: being pitches, it results in a sequence of major thirds. The starting pitch is very low (20, less than 26

Hz). Each routine repeatedly generates an event (a square wave that is percussively enveloped). The generation interval is given by the geometric series times. Pitch, repetition interval and stereo panning are linked to the counter (it is the so-called *parameter linking*), so that low pitches are more frequent and are placed on the left, while high pitches are less frequent and placed more on the right. The effect is a sort of crumbled arpeggio that produces a progressive dephasing, before returning to minimum duration given by the common multiple of durations. The delay in the attacks depends on the fact that first the routines wait (a variable time), and then a synth is created.

```
1  (
2  var num = 20 ;
3  var pitches = Array.series(num, 0, 4)+20 ;
4  var times =  Array.geom(num, 1, 1.01) ;
5  num.do{|i|
6      { inf.do{
7          times[i].wait;
8          {
9              Pan2.ar(
10                 in:    Pulse.ar(pitches[i].midicps)
11                 *
12                 EnvGen.kr(Env.perc, 1, 0.2, doneAction:2),
13                 pos: i.linlin(0, num.size-1, -1, 1)
14             )
15         }.play
16      }}.fork
17 } ;
18 s.scope ;
19 )
```

7.6 Graphic interlude: drawings and animations

SuperCollider provides great graphical capabilities, not only within the set of objects that directly implement user interaction (e.g. buttons, sliders, knobs, and so on), but also in relation to generative graphics. The next example illustrates some basic aspects.

```
 1  w = Window("tester", Rect(10, 10, 500, 500))
 2  .background_(Color.white).front ;
 3  w.drawFunc = {
 4      Pen.strokeColor = Color.rand(0.1, 0.9) ;
 5      Pen.moveTo(0@250) ;
 6      Pen.lineTo(250@250) ;
 7      Pen.lineTo(500@500) ;
 8      Pen.stroke ;
 9  } ;

11  w.refresh ;
```

As shown, a window (the same also applies to the graphical class that is specifically designed as a "canvas", UserView) includes, among its methods drawFunc, literally, a drawing function (3-9). This method is evaluated when the window is built (if defined at initialization), but can be called on demand by the method refresh (11). By evaluating again line 11, the color of the line changes, as it depends on a randomly chosen color (line 4). Since the choice is random inside the function, any new evaluation, that can be forced with re-fresh, actually results in a new color (or even better, drawing a new line with a random color). Inside drawFunc, and only there, it is possible to use the class Pen, which does not have instance methods but only class ones: very simply, the user does not create Pen-like objects, rather s/he directly uses the methods provided by the class. Pen implements a working mode that is typical of many graphic programming languages (primarily PostScript, but also Processing or NodeBox), and that stems directly from the control of mechanical plotters, that were provided with a mechanical arm. This detail is not inessential to understand the way in which Pen works. Basically, Pen is like a pen that can be asked to move to a certain position, to draw a line, change the color of the ink, to fill an area, to draw an oval, and so on. In this example the stroking color is first defined (strokeColor, 4); then the pen is moved to the point $(0, 250)$. Two observations: the syntax 0@250 is an abbreviation ("syntactic sugar") for constructing an instance of the class Point, i.e. it is the same as Point (0,250). Here, Point intuitively implements the geometrical notion of a point in a plane; move does not indicate tracing a line, rather to move the pen to the specified position (it is like moving the arm in the plotter without drawing). Then two lines are drawn through the method lineTo. Note that the process is incremental, that is, the pen moves to a point, where it remains until further notice: the point is its starting position. Until this moment, we have provided a description of what the plotter Pen should do, but the action is performed only when the stroke method is invoked on Pen. Only at that point, the instructions are actually taken over by the drawing process. In short, Pen defines a finite state machine for drawing, and a very efficient one.

The next example shows how to interact with Pen.

```
1  // circle dimensions
2  ~dim = 50 ; ~dim2 = 25 ;
3  // color parameters
4  ~hue = 0; ~sat = 0.7; ~val = 0.7;

6  w = Window("tester", Rect(10, 10, 500, 500))
7  .background_(Color.white).front ;
8  w.drawFunc = {
9      var oo = 250-(~dim*0.5) ; // to place the circle in the middle
10     Pen.addOval(Rect(oo, oo, ~dim, ~dim)) ; // the circle
11     Pen.color_(Color.hsv(~hue, ~sat, ~val)) ; // color
12     Pen.fill ; // fill it
13     oo = 250-(~dim2*0.5) ; //  to place the circle in the middle
14     Pen.addOval(Rect(oo, oo, ~dim2, ~dim2)) ; // the circle
15     Pen.color_(Color.white) ; // color
16     Pen.fill ; // fill it
17 } ;

19 ~hue = 0.2; w.refresh ;
20 ~dim = 200; w.refresh ;
21 ~dim2 = 10; w.refresh ;
```

The function assigned to drawFunc draws a circle using the method addOval. The method draws an oval by defininig the rectangle in which it is included: or, in the case of a square, the result would be a circle (10). The oval's side is given by the environmental variable ~dim. The circle is centered in the window in line 9 (the square is in fact defined by its top left vertex, and therefore it must be shifted appropriately). Then, the circle is colored: first the color of the pen is defined (11) by three environmental variables(4), then the pen is asked to fill the surface with the method fill (note: not with stroke, 12). In the next lines, the process is repeated by adding a second circle with a size ~dim2 and filled with the background color (white), thus producing a "hole" effect as it overlaps the previous circle. Remember that Pen is incremental: if ~dim2 would be greater than ~dim, the second circle would cover the first, and nothing would be visible.

Lines 19-21 show the interaction made possible by the environmental variables. It is possible to change their values and to force the window to upgrade by calling drawFunc through refresh. Pen offers many possibilities, the exploration of which is outside the scope of this text and it is left to the reader, who can discover more from the help file. Also, as it is possible to trace the position of the mouse, it is thus possible to visually interact with the generative graphics. But what is mostly relevant in this chapter is scheduling applied to graphics (that is, animation). The following code, which assumes that the former has been evaluated, is now easy to understand:

```
1  (
2  {inf.do{|i|
3      k = i*3 ;
4      ~hue = k%500/500;
5      ~dim = k%500;
6      ~dim2 = ~dim*0.25 ;
7      w.refresh ;
8      0.1.wait ;
9  }}.fork(AppClock)
10 )
```

An infinite routine uses the counter i to calculate a second counter that increases at a faster rate, k. So, the variable associated with the color, ~hue, is obtained by the modulo 500 on the counter, and divided by 500 (4). The rationale is that the maximum value must be 1, which is the maximum allowed for the argument hue in Color.hsv. Similarly, ~dim is applied the same module (5), so that it does not exceed the window size (a fact that per se does not throw an error, but that evidently is not desired). Then, dim2~ is calculated as $\frac{1}{4}$ (0.25, 6) of ~dim. The window is updated and the operation is repeated after 100 ms. Finally, the routine is scheduled on AppClock (otherwise, defer should be used).

7.7 Routines vs. Tasks

Routines can be reset in the initial condition through the message reset and ended by the message stop. However, when they receive the message stop, in order to start again, they must first receive the message reset that resets them to the starting condition. This aspect represents a potentially important limitation in their musical use. The class Task implements a process "that can be paused" (a *pauseable process*). In terms of implementation, the behavior of a task is very similar to that of a routine, as the following example demonstrates:

```
1  (
2  t = Task({
3          inf.do({|i| i.post; " steps toward nothing".postln; 1.wait})
4      }) ;
5  )

7  // start
8  t.play ;

10 // pause: internal state is stored
11 t.pause ;

13 // restart from last state
14 t.resume ;

16 // restart from state 0
17 t.reset ;

19 // stop: same as pause
20 t.stop ;

22 // play: same as resume
23 t.play ;
```

As we see, the methods stop/pause, and play/resume behave exactly the same way. Looking in detail, a routine is –in certain conditions– slightly more accurate in case of stop/restart. In any case, as a general rule, if a process does not have to be controlled (because it ends by itself, or it never ends at all) a routine should be used; if it has to be stopped and resumed, then a task is the right choice. The following example uses a task to "pause" a process. The SynthDef "bink" also gives us a chance to discuss some aspects of signal processing.

```
 1  (
 2  SynthDef("bink", { arg freq = 440, pan = 1;
 3      var sig, del;
 4      // source
 5      sig = Pulse.ar(freq
 6          *Line.kr(1,
 7              LFNoise1.kr(0.1)
 8              .linlin(-1,1, -0.5, 0).midiratio, 0.1),
 9          width:0.1
10      ) ;
11      // delay tail
12      del = Mix.fill(20, {|i|
13          DelayL.ar(sig,
14              delaytime: LFNoise1.kr(0.1)
15              .linlin(-1,1, 0.01, 0.1)
16          )
17      }) ;
18      // mix, envelope, spatialization
19      Out.ar(0,
20          Pan2.ar(
21              (sig+del)*EnvGen.kr(Env.perc, doneAction:2),
22              LFNoise1.kr(pan), 0.1
23          ))
24  }).add;
25  )
26  s.scope ;
27  x = Synth(\bink ) ;
28  x = Synth(\bink , [\freq , 60.midicps]) ;
```

The SynthDef is logically divided into 3 blocks. In the first one, a sound source is defined as a "tight" square wave, that is, with a reduced duty cycle (width: 0.1). Its base frequency freq is multiplied by a ramp signal that goes from 1 (therefore, no change) to a pseudo-random value in the range $[-0.5, 0]$, expressing a "detuning" in semitones. In fact, midiratio converts a value in semitones into a multiplier for frequency: 0.midiratio indicates 0 semitones and is equal to 1, 12.midiratio indicates 12 semitones and is equal to 2^7. The use of LFNoise1 ensures that the transition between the consecutive generated values is continuous. In essence, the result is a glissando decreasing from the

---

[7] Because of approximations in floating point notation, once evaluated the expression prints 1.9999999999945, but this is quite normal.

frequency `freq`, for a maximum of a quarter tone (half a semitone). The second block (11) uses the UGen `Delay` to create copies of the delayed signal through the argument `delaytime`, that intuitively defines how much the signal has to be delayed. The delayed signal is again managed in a pseudo-random way by means of `LFNoise1`, the output of which oscillates in a range of $[10, 100]$ ms. The delay is placed inside `Mix.fill`, which then produces 20 copies with pseudo-random delayed signals whilst mixing them. In the third block (19), the non delayed copy is mixed with the delayed signal and enveloped by a percussive envelope (it is a "note"-like synth). At the end, the whole is distributed on the stereo front with a pseudo-random panning. The obtained sound ressembles a kind of plucked metal string tone, where the metallic feature is provided by the full spectrum of the square wave, the percussive pizzicato by the envelope, and the (vague) physical simulation effect by the presence of delays, which in some way simulates an acoustic resonator, as the latter behaves like a kind of small echo room. The following code defines a set of processes that exploit the SynthDef.

```
1  (
2  var arr, window ;
3  arr = Array.fill(8, { arg i ;
4       Task({
5            var pitch ;
6            inf.do({
7                 pitch = (i*[3, 5, 7].choose+40)%80+20 ;
8                 Synth(\bink , [\freq , pitch.midicps,
9                     \pan , pitch/50 ]) ;
10                ((9-i)/8).wait ;
11           })
12       })
13 }) ;

15 arr.do({ |t| t.play}) ; // play all

17 // GUI for tasks
18 window = Window("control", Rect(100, 100, 8*50, 50)).front ;
19 8.do {|i|
20      Button(window, Rect(i*50, 0, 50, 50))
21      .states_([[i.asString, Color.red],[i.asString, Color.black]])
22      .action_{|me| if(me.value == 1) {arr[i].pause}{arr[i].play}}

24 }
25 )
```

The variable arr is assigned an array that contains 8 tasks. Each of them generates a voice in which the pitches depend on the counter (in substance the voices are differentiated in relation to register). The task selects a value between [3, 5, 7], multiplies it by the index (the voice register, so to say). The starting point is the pitch 40, and the progression is blocked to a value of 80. The pitch sequence in each layer is complex because it depends on the relationship between the counter i, the array [3,5,7] and the modulo operator 20. Each layer has its own autonomous development rate: the lower ones move slower, the higher one faster, in the range of [0.251, 125] seconds (respectively i = 7 and i = 0). Note the quantization to a sixteenth note at bpm = 60. The progression proceeds inexorably (15) in parallel. The next block (17-24) allows one to interactively explore the layers and to vary their relative phase. A window contains 8 two-state buttons. Quite simply, by pressing them, the relative task is dis/activated. This makes it possible to listen to the individual layers and also change their timing.

7.8 Patterns

It is quite clear that routines and tasks offer a greater flexibility in time control. But SuperCollider also makes available certain types of data structures that, while not necessarily related to scheduling, also find their most typical application in the management of events in time, Patterns. In some sense, such an organization of information has already been discussed in the context of the Demand UGens. Better said, Demand UGens implement on the server side some features that Patterns define on the client side. However, the latter are in reality more often used.

Let us reconsider what has been discussed above. A pattern is the form, described procedurally, of a data sequence. It is not the sequence that is described as such, rather a rule for its production. For example, Pseq describes sequences obtained by repeating a certain sequence: so, Pseq ([1,2,3,4], 3) describes a sequence obtained by repeating three times the four values contained in [1,2,3,4]. The following code:

```
1 p = Pseq([1,2,3,4], 3).asStream ;
2 13.do{ "next...? -> ".post; p.next.postln } ;
```

Once evaluated, prints on the post window:

```
 1 next...? -> 1
 2 next...? -> 2
 3 next...? -> 3
 4 next...? -> 4
 5 next...? -> 1
 6 next...? -> 2
 7 next...? -> 3
 8 next...? -> 4
 9 next...? -> 1
10 next...? -> 2
11 next...? -> 3
12 next...? -> 4
13 next...? -> nil
14 13
```

The previous code assigns the variable p a Pattern Pseq, but not only: through
the message asStream, it generates the actual sequence. In other terms, Pseq is a
form that, in order to be used, must first be converted into a stream (stream) of
data. Note that the message asStream returns a routine. In fact, on the routine
it is possible to call the method next for 13 times, so as to access the next value:
when the sequence is over (the thirteenth time), next returns nil .

The next SynthDef computes a signal sig by summing 10 sinusoids, their
frequencies varying by (a maximum of) 2.5% (3): the result is a sort of chorus
effect, with a slight vibrato. A reverberated version is enveloped percussively
(4). To let the reverb develop fully, the SynthDef relies on the UGen DetectSi-
lence (5). This UGen analyzes the input signal and when the amplitude of the
latter falls below a threshold, executes doneAction. The UGen returns a binary
signal, 0 or 1 relative to exceeding the threshold. This value can be used to an
advantage as a trigger. This is not the case of the example. Here, DetectSi-
lence verifies that, once the signal is terminated (i.e. its amplitude falls below
a threshold), the synth is automatically deallocated through doneAction: 2. In

this way, it is not necessary to use a bus to route the signal to a reverb unit synth in order to have a complete reverberation. The signal sig is spatialized as usual on two channels (6).

```
1  SynthDef(\sinPerc , { |freq = 440, pos = 0, level = 0.125, detune = 0.025|
2      var sig =
3      Mix.fill(10, {SinOsc.ar(freq+Rand(0, freq*detune))}) ;
4      sig = FreeVerb.ar(sig* EnvGen.kr(Env.perc)) ;
5      DetectSilence.ar(sig, -96.dbamp, doneAction:2) ;
6      Out.ar(0, Pan2.ar(sig, pos, level))
7  }).add ;
```

The next example includes three processes that use the pattern Pseq. In the first, a sequence of 10 pitches is coded by intervals (in semitones) from a base ($= 0$). The routine (4-9) generates a set of notes every $\frac{1}{4}$ of a second, starting from p, adding 70, and converting them into frequencies. The second block (12-22) uses the same sequence, but it defines another sequence for durations, expressed in units (15) and later scaled (19). Since the sequence of pitches has a different size from that of durations (respectively, 10 vs 9), it implements a classic situation, the dephasing process known as *talea* vs *color*. The two systems will be in phase again after $10 \times 9 = 90$ events. The third case is much more complex to look at, but it simply extends the same principle of asymmetry *talea/color* to other dimensions. In particular, it defines a pattern that controls density, ~density: an event can include from 1 to 4 parallel synths (i.e. ranging from a note to a tetrachord). Each note is then harmonized according to the pattern ~interval that defines an interval starting from the departing pitch. Finally, the calculation of the actual pitch takes into account ~octave, that shifts the value in relation to the octaves. Note also that the amplitude is scaled according to the density, so that chords with different densities will all have the same amplitude.

```
1  (
2  // 1. a sequence
3  p = Pseq([0,2,4,3,10,13,12,6,5,7], inf).asStream ;
4  {
5      inf.do{
6          Synth(\sinPerc , [\freq , (p.next+70).midicps]) ;
7          0.25.wait ;
8      }
9  }.fork
10 )

12 (
13 // 2. color vs. talea (10 vs 9)
14 p = Pseq([0,2,4,3,10,13,12,6,5,7], inf).asStream ;
15 q = Pseq([1,1,2,1,4,1,3,3,1], inf).asStream ;
16 {
17     inf.do{
18         Synth(\sinPerc , [\freq , (p.next+70).midicps]) ;
19         (q.next*0.125).wait ;
20     }
21 }.fork
22 )

24 (
25 // 3. color vs. talea (10 vs 9)
26 // vs. chord dimension vs. interval vs. octave...
27 p = Pseq([0,2,4,3,10,13,12,6,5,7], inf).asStream ;
28 q = Pseq([1,1,2,1,4,1,3,3,1], inf).asStream ;
29 ~density = Pseq([1,2,4,1,3], inf).asStream ;
30 ~interval = Pseq([3,4,7,6], inf).asStream ;
31 ~octave = Pseq([-1,0,0,1], inf).asStream ;
32 {
33     inf.do{
34         var den = ~density.next ;
35         den.do{
36             var delta = ~interval.next ;
37             var oct = ~octave.next ;
38             Synth(\sinPerc ,
39                 [
40                     \freq , (p.next+70+delta+(12*oct)).midicps,
41                     \level , 0.1/den
42             ]) } ;
43         (q.next*0.125).wait ;
44     }
45 }.fork
46 )
```

A canon structure is at the basis of the next program, which defines a sequence of pitches, durations and panning positions (2-3). It builds 3 arrays of 4 routines, as indicated by Pseq().asStream. Each array contains the control data for a musical sequence. The next cycle starts 4 parallel routines. Each routine refers by means of the counter i to the relative element in the routine. In addition, a multiplier for pitches and durations is calculated from i. The result is a four voice canon in different octaves.

```
1  (
2  var mel = [0,3,5,6,7,9,10], rhy = [1,1,3,2,1,2] ;
3  var pan = Array.series(5, -1, 2/4) ;
4  var arrPseq = Array.fill(4, { Pseq(mel, inf).asStream } ) ;
5  var durPseq = Array.fill(4, { Pseq(rhy, inf).asStream } ) ;
6  var panPseq = Array.fill(4, { Pseq(pan, inf).asStream } ) ;

8  4.do{|i|
9      { inf.do{
10          var freqSeq = arrPseq[i] ;
11          var freq = (12*i+freqSeq.next+48).midicps ;
12          var durSeq = durPseq[i] ;
13          var dur = durSeq.next*0.125*(i+1) ;
14          var pan = panPseq[i].next ;
15          Synth(\sinPerc , [\freq , freq, \pos , pan, \level , 0.07]) ;
16          dur.wait ;
17      }
18      }.fork
19  }
20  )
```

Two more examples of algorithmic composition. The next uses three SynthDefs:

```
1  SynthDef(\sinPerc , {
2      |out = 0, freq = 440, pos = 0, level = 0.125, detune = 0.025|
3      var sig =
4      Mix.fill(10, {SinOsc.ar(freq+Rand(0, freq*detune))}) ;
5      sig = FreeVerb.ar(sig* EnvGen.kr(Env.perc)) ;
6      DetectSilence.ar(sig, -96.dbamp, doneAction:2) ;
7      Out.ar(out, Pan2.ar(sig, pos, level))
8  }).add ;

10 SynthDef(\impPerc , {
11     |out = 0, freq = 440, pos = 0, level = 0.125, detune = 0.025|
12     var sig =
13     Mix.fill(10, {Impulse.ar(freq+Rand(0, freq*detune))}) ;
14     sig = FreeVerb.ar(sig* EnvGen.kr(Env.perc)) ;
15     DetectSilence.ar(sig, -96.dbamp, doneAction:2) ;
16     Out.ar(out, Pan2.ar(sig, pos, level))
17 }).add ;

19 SynthDef(\pulsePerc , {
20     |out = 0, freq = 440, pos = 0, level = 0.125, detune = 0.025|
21     var sig =
22     Mix.fill(10, {Pulse.ar(freq+Rand(0, freq*detune), width:0.1)}) ;
23     sig = FreeVerb.ar(sig* EnvGen.kr(Env.perc)) ;
24     DetectSilence.ar(sig, -96.dbamp, doneAction:2) ;
25     Out.ar(out, Pan2.ar(sig, pos, level))
26 }).add ;
```

that are used in the following process:

```
1  (
2  ~bs = Bus.audio(s, 2) ;

4  a = Pseq([1,3,4,1,1], 2) ;
5  b = Pseq([1,1,1]/3, 1) ;
6  c = Pseq([1,1,2]/2, 2) ;
7  d = Pseq([-1], 1);
8  e = Prand([5,0], 3);

10 f = Pseq([0,0, 1,0,0,1,2, 2,0,1,2,0,1,2], inf).asStream ;

12 p = Pxrand([a,b,c, d,e], inf).asStream ;

14 {
15     inf.do{
16         var which, id ;
17         n = p.next ;
18         if (n == -1) {
19             [20,40].do{|i|
20                 Synth(\sinPerc ,
21                     [\freq , i.midicps, \detune , 0.0125, \level , 0.2,
22                         \out , ~bs]);
23             } ;
24             n = 1/8 ;
25         } {
26             id = f.next ;
27             which = [\sinPerc , \impPerc ,\pulsePerc ][id] ;
28             Synth(which,
29                 [\freq , (n*3+70).midicps,
30                     \detune , 0.05, \out , ~bs, \pos , id-1*0.75]);
31             Synth(which, [\freq , (n*3+46).midicps,
32                 \detune , 0.025, \out , ~bs, \pos , id-1*0.75]);
33         } ;
34         (n*0.25*60/84).wait
35     }
36 }.fork ;

38 x = {|vol = 0.5| Out.ar(0, In.ar(~bs, 2)*vol)}.play(addAction:\addToTail ) ;
39 s.scope ;
40 x.set(\vol , 0.15)
41 )
```

The details are left to the reader. It is worth to highlight three aspects. First, the presence of a conditional within the routine that takes into account the value of p.next. If the value is −1 then the first block is executed, which then redefines the value of n, overwriting the one initially assigned through p.next. Secondly, note that the SynthDef in the second block is chosen from a pattern (f). Finally, to control the amplitude of the overall signal in a effective and efficient way, all the signals are written on the bus ~bs. The bus content is routed into the synth x which has a function to scale the overall signal amplitude before sending it out.

The following example uses the SynthDef sinPerc to generate a set of arpeggios that are vaguely reminiscent of an electric piano or celesta. The first block defines the control data structures. In particular, the array ~arr contains a set of patterns intended to handle the harmonization (i.e. the pitches composing the arpeggio). The array p contains a base pitch that will be added to 60 (21). The array q contains the values that define the speed of the arpeggio, while the time between arpeggios is set by timebase (31). Note that this value is related to the parameter oct that defines the octave transposition for the pitch base by means of the control structure case (17-20). Lower arpeggios are played more slowly than the higher ones.

```
1  (
2  a = Pseq([3,4,4,3,3], inf) ;
3  b = Pseq([6,7,7], inf) ;
4  c = Pseq([9,10, 9, 9], inf) ;
5  d = Pseq([11,12,13,12,14], inf) ;
6  e = Pseq([5, 6,5]+12, inf) ;
7  f = Pseq([8,9,9,8,9]+12, inf) ;

9  ~arr = [a,b,c,d,e,f].collect{|i| i.asStream} ;
10 p = Pseq([0,3,3,2,2, 3,0, 7, 6, 11,11, 10, 9, 12], inf).asStream ;
11 q = Prand([ 1/64, 1/32, 1/16], inf).asStream ;
12 r = Pseq(
13      Array.fill(12, {0})++
14      Array.fill(3, {-12})++
15      Array.fill(8, {7}),
16      inf).asStream ;
17 )
18 (
19 var base, harm, timeBase, oct ;
20 {
21      inf.do{
22          oct = r.next ;
23          case
24          {oct == 7} { timeBase = 1/16}
25          {oct == -12} { timeBase = 2/3}
26          {oct == 0} { timeBase = 1/4} ;
27          base = p.next+60+oct ;
28          Synth(\sinPerc ,
29              [\freq , base.midicps, \detune , 0.01,\level , -30.dbamp]) ;
30          6.do{|i|
31              harm = base+~arr[i].next ;
32              Synth(\sinPerc ,
33                  [\freq , harm.midicps, \detune , 0.01,
34                      \level ,-30.dbamp, \pos , i.linlin(0,1.0, -0.5,0.5)]) ;
35          q.next.wait;
36          } ;
37          timeBase.wait ;

39 }}.fork ;
40 )
```

There are many patterns defined in SuperCollider, not only Pseq and Pxrand: for example, some are also designed to manage the audio, not just the control of events, while others allow one to filter the content of data streams generated by other patterns (the so-called *filter patterns*). It could be said that patterns are a sort of sublanguage of SC specifically dedicated to the representation of data streams.

7.9 Events and Event patterns

Patterns are data structures that encode in compact form data sequences. As seen in the previous examples, they can be used within the usual scheduling processes, that is as, routines or tasks. They are however also a vital component of a very peculiar logic of sequencing, which is based on the concept of "event". For example, the following mysterious expression generates sound:

```
1  ().play ;
```

Empty brackets are an abbreviation for the creation of an instance of the class Event, and therefore the expression is the same as

```
1  Event.new.play ;
```

An "event" in SC is simply an association between environmental variables and values, which can respond to the method play. Technically, it is in fact nothing more than a dictionary mapping names to values. The class Event predefines many of these names, as well as functions that are associated with those names, in order to specify their semantics. If we evaluate the above line, two things happen, we hear a sound and we see on the post window something like the following:

```
1  ( 'instrument': default, 'msgFunc': a Function, 'amp': 0.1, 'sustain': 0.8,
2  'server': localhost, 'isPlaying': true, 'freq': 261.6255653006,
3  'hasGate': true, 'detunedFreq': 261.6255653006, 'id': [ 1002 ] )
```

It is possible to recognize names typically related to synthesis, and associated with values. This is because SC defines a default event in which a synth built from the SynthDef \default (variable 'instrument')[8] plays a 261.6255653006 Hz tone ('freq') with an amplitude 0.1 ('amp'), and with other parameters. This event is created when the method play is run. The following example makes the situation clearer:

```
1  (\instrument : \sinPerc , \midinote : 65).play ;
```

Here the instrument is instead \sinPerc (the SynthDef from previous examples) while the frequency is not explicitly defined with a reference to 'freq' but by setting 'midinote': this variable is internally associated with a function that automatically defines the relative value of 'freq' (note the post window). Two considerations are important:

1. Event predefines many variables. In this way, it is possible to manage in a very simple way various pitch systems (scales, modes, and not necessarily equally temperate tunings) without having to calculate individual frequencies. The same applies for other parameters. The reader is invited to read the relative help files;
2. in order to use the variables in relation to a SynthDef, the latter must be equipped with arguments with the variables' names. For example, an event is considered as a note, so it is necessary to use doneAction: 2 in the SynthDef, otherwise the instantiated synth will not be deallocated. Still, some names –such as freq, amp, gate– are conventionally 'in use' with Event, and must be respected in the SynthDef provided by the user.

---

[8] The SynthDef \default is loaded on the server on booting.

In the next example, a SynthDef includes freq but not doneAction: 2. Pitches are correct (freq is properly calculated from the variable midinote), but the event does not end, as every synth remains active.

```
1  SynthDef(\cont , {arg freq = 440; Out.ar(0, SinOsc.ar(freq))}).add ;

3  (\instrument : \cont , \midinote : 80).play ;
4  (\instrument : \cont , \midinote : 60).play ;
```

In the next example the concept of the event is used in conjunction with a Pattern.

```
1  (
2  SynthDef(\sinPerc , { |freq = 440, pos = 0, amp = 1, detune = 0.025|
3      var sig =
4      Mix.fill(10, {SinOsc.ar(freq+Rand(0, freq*detune))*0.1}) ;
5      sig = FreeVerb.ar(sig* EnvGen.kr(Env.perc)) ;
6      DetectSilence.ar(sig, -96.dbamp, doneAction:2) ;
7      Out.ar(0, Pan2.ar(sig, pos, amp))
8  }).add ;
9  )

11 p = Pseq([0,2,5,7,9], inf).asStream ;

13 {
14     inf.do{
15         (\instrument :\sinPerc , \amp :0.5, \ctranspose :p.next).play ;
16         0.25.wait ;
17     }
18 }.fork ;
```

First, the SynthDef 'sinPerc' is rewritten so as to comply with the model provided by Event. The amplitude is normalized so that the mix at most results in a peak value = 1 (4), and the argument level is replaced by amp. The pattern p represents a pentatonic melody (11). The next routine instantiates each 250 ms (16) a new event using sinPerc, defines the argument amp and passes the next value of p to the variable \ctranspose, yet another model for the definition of pitches, that adds the passed value to midi base note (by default = 60) and converts it into the relative frequency.

It is perfectly legitimate to use the concept of event in this way, but the most common use (and in some way the rationale underlying the concept) is linked to "event patterns". The patterns discussed so far are usually described as a "value" pattern or "based on the list". Event patterns provides further possibilities. An event pattern, typically Pbind, associates events with data patterns, which can then be executed. This results in a compact and elegant notation for the specification of streams of sound events. In the minimal example that follows, first an event pattern p is defined (as in the example above); then, an event pattern that associates the variable \ctranspose with the pattern p (3) and the other variables with the values described in the previous example. Finally, play is invoked: it returns an object of type EventStreamPlayer, literally a performer of event streams, a kind of player that generates sound from given specifications. The player, f (5), can be controlled interactively in real-time (6-9). Note that it is the EventStreamPlayer that takes care of generating streams from the pattern.

```
1 p = Pseq([0,2,5,7,9], inf) ;

3 e = Pbind(\ctranspose ,p,\instrument ,\sinPerc ,\amp ,0.5,\dur ,0.25) ;

5 f = e.play ;
6 f.pause ;
7 f.play ;
8 f.mute ;
9 f.unmute ;
```

The next example introduces two patterns p and d for pitches and durations, following the model *talea/color* (1-3). A single sequence can be played with the block 2 (1-5). Block 3 builds a polyphony of 4 voices in different octaves and wherein pitch is proportional to the multiplier for the dynamic range (17-18). Note that line 21 reassigns to a (which contained the first instances of Pbind)) their relative EvenStreamPlayers. If block 3 is still running, it is possible to evaluate the block 4, this starts routines that pause and then restore the layers (once per second). Finally, block 5 stops the routine r and pauses all the EventStreamPlayer.

```
 1 // 1. pattern
 2 p = Pseq([0,2,3,  5,  6,  0,2,3,  5,  7,  8,0,7], inf);
 3 d = Pseq([1,1,2,1,1,2,3,1,1,2], inf);

 5 // 2. first test, talea vs. color
 6 (
 7 Pbind(
 8     \instrument , \sinPerc , \amp , 0.25,
 9     \ctranspose , p, \dur , d*0.125).play ;
10 )

12 // 3. canon
13 (
14 a = Array.fill(4, {|i|
15     Pbind(
16         \instrument , \sinPerc , \amp , 0.25,
17         \root , i*12-24,
18         \ctranspose , p, \dur , d*0.125*(i+1))
19 }) ;

21 a = a.collect{|i| i.play} ; // a now contains the players
22 )

24 // 4. a process that pauses selectively
25 (
26 r = {
27     inf.do{|i|
28         {b = a[i%4].mute ;
29             2.wait ;
30             b.unmute ;}.fork;
31         1.wait ;
32     }
33 }.fork
34 )

36 // 5. all paused
37 r.stop; a = a.do{|i| i.pause} ;
```

It may not always be intuitive to interact with event patterns, because in some way the environmental variables that are organized in the events always require us to operate within a specific "pattern logic". To this purpose, the following example demonstrates the use of Pfunc, a pattern that, at each call, returns a value calculated by a function. In this situation, the function is used to access the value i from the array r, an operation that allows us to define for each event pattern a different transposition. Line 16 then defines a sequence [-12, -5, 2, 9, 16] while line 17 restores the original sequence [0, 0, 0, 0, 0], in which the four voices sound a canon in unison.

```
1  Pbind(\ctranspose , p, \stretch , 1/2, \dur , d, \root , 1).play ;

3  ~seq = [0,2,5,7,9] ;
4  p = Pseq(~seq, inf) ;
5  d = Pseq([1,1,2,1,1,3,1,4,1,1], inf) ;
6  r = Array.series(~seq.size, 0, 0) ;

8  (
9  8.do{|i|
10     Pbind(\ctranspose , p+Pfunc({r[i]}), \stretch , 1/( i+1), \dur , d,
11     ).play ;
12 } ;
13 )

15 // control
16 r = Array.series(~seq.size,-12,7) ;
17 r = Array.series(~seq.size,0,0) ;
```

The last example also intends to highlight the expressive potential arising from pattern nesting. The SynthDef is a slightly modified version of a previous one. In particular, we have added the argument amp and panning which are now no longer run by a pseudo-random UGen.

```
 1  SynthDef("bink", { arg freq = 440, pan = 0, amp = 0.1;
 2      var sig, del;
 3      // source
 4      sig = Pulse.ar(freq
 5          *Line.kr(1,
 6              LFNoise1.kr(0.1)
 7                  .linlin(-1,1, -0.5, 0).midiratio, 0.1),
 8          width:0.1
 9      ) ;
10      // delay tail
11      del = Mix.fill(20, {|i|
12          DelayL.ar(sig,
13              delaytime: LFNoise1.kr(0.1)
14                  .linlin(-1,1, 0.01, 0.1)
15          )
16      }) ;
17      // mix, envelope, spatialization
18      Out.ar(0,
19          Pan2.ar(
20              (sig+del)*EnvGen.kr(Env.perc, doneAction:2),
21              pan, amp
22      ))
23  }).add;
```

The first block (1-24) organizes durations. The extensive, messy use of environment variables was left as an "ethnographic" evidence of the incremental and interactive composition work. The basic idea in relation to time organization is to divide the process into sections. The first section presents two variants; a "fill-like" section follows, that selects between a set of fairly extensive rhythmic possibilities. As we see, durations are specified using integers, that ensure a simpler management. The two patterns a and b last both 24 units, and the first of them alternate the durations of 7 and 5 units for the pattern. The sections "fill" (7, 15) always last 16 units and organize rhythmically denser patterns (e.g. note the durations). Lines 17 and 18 define two pattern Prand that choose at random one of the phrases (h) and one of the fill (i). Thus, the pattern i is an infinite sequence of h and i, that is, a phrase with a fill. The lower voice is made up of three "measures" of a duration of 40 units (j), plus a specific sequence of pitches s. This pattern also shows the rhythmically denser special use of the \r to indicate a pause. The two Pbind use \stretch as an indication of tempo (defined empirically). The bass voice ~e2 proceeds in an ordinary way. The higher voice ~e1 uses Pkey, a special pattern that allows to access the value of another variable in Pbind. Midi notes to be played are defined in this way: through a linear interpolation (a mapping), from durations to pitches (thus, longer durations result in lower pitches, and vice versa).

```
1  (
2  //  duration organisation
3  // phrase
4  a = Pseq([7,5], 2) ;
5  b = Pseq([8], 3) ;
6  // fills
7  c = Pseq([4], 4) ;
8  d = Pseq([16/3], 3) ;
9  e = Pseq([1,2,3,4,6], 1) ;
10 f = Pseq([8/3, 16/3], 2) ;
11 g = Pseq([6,5,4,1], 1) ;
12 y = Pseq([5,5,1,5], 1) ;
13 z = Pseq([4,3,2,5,2], 1) ;
14 w = Pseq([30/5, 25/5, 15/5, 10/5], 1) ;
15 u = Pseq([30/5, 25/5, 15/5, 10/5].reverse, 1) ;
16 // first nesting
17 h = Prand([a, b],1) ;
18 i = Prand([c,d,e, f,g, y,z,w,u], 1) ;
19 // second
20 k = Pseq([h, i], inf) ;
21 // a bass line
22 j = Pseq([10, 20, 10, 40, 40/3, 40/3, 40/3], inf) ;
23 l = Pseq([
24     -24, -24, -12, -24, \r , -26, -25,
25     -24, -14, -12, -24, -13, \r , -25,
26 ], inf) ;
27 )

29 (
30 // two parallel voices
31 ~e1 = Pbind(\instrument , \bink , \stretch , 60/32/60, \dur , k.next,
32     \ctranspose , Pkey(\dur ).linlin(0, 12, 12, 0.0), \pan , rrand(-0.5, 0.5)
33 ).play ;

35 ~e2 = Pbind(\instrument , \bink , \ctranspose , l.next, \amp , 0.1,
36     \stretch , 60/32/60, \dur , j.next).play
37 )
38 // stop
39 [~e1, ~e2].do{|i| i.pause} ;
```

7.10 Conclusions

Time is obviously a key issue in the management of sound and this chapter was simply an attempt to suggest some early indications about the many possibilities (conceptual and operational) that SuperCollider provides. This expressive richness makes SC suitable for many different situations, benefiting from specific modes of conceptualization. However, what has been discussed is only, one could say, the visible part of a very deep iceberg, which the reader is invited to explore through the help files. The constant interaction between programming and sound output that SC provides makes the task extremely inspiring.

8 Synthesis, II: introduction to basic real-time techniques

So far, audio synthesis has been introduced with respect to acoustics and to control signals. The next chapter will focus on real-time audio synthesis techniques[1] in regard to SuperCollider's implementation.

In the following we will extensively use the syntax shortcuts for func.play (or scope, freqscope, plot), which are compact and particularly useful when experimenting with UGen graphs for audio processing/synthesis; amongst other things they are also abundantly used in the help files of the various UGens. In the few examples that involve SynthDefs, the latter will not be particularly complicated, given that the focus here is on audio synthesis techniques themselves rather than their optimal implementation.

8.1 Oscillators and tables

A sound synthesis algorithm is a formalized procedure with the purpose of generating a stream of numbers that represent an audio signal. When non-real-time sound synthesis is concerned, the signal generation could begin from just a mathematical function such as $sin(x)$. Albeit conceptually elegant, such an approach would be highly inefficient from a computational point of view in

---

[1] The chapter closely follows the approach proposed in *Audio and multimedia*, which the interested reader may refer to.

real-time audio synthesis: it would cause the CPU to calculate such a function as many times per second as that of the working sampling rate chosen (i.e.44.100 which is the default sampling rate in SC). It becomes immediately clear that the audio synthesis also revolves around the choice of those particular algorithmic strategies that achieve the desired effect while also being computational "cheap". Regarding the construction of a periodic waveform, there is another method, one with a rather long tradition in computer music: building a *digital oscillator*. Digital oscillators are fundamental algorithms in computer music since they are both used in order to generate audio/control signals directly, as well as within the context of more complex sound generators. As far as a sinusoidal signal is concerned (and in general, as far as the stationary part of every periodic sound is concerned), it is predicable to a great extent: it typically repeats the very same values at each period. Such a period can be easily represented in the form of sampled values stored equidistantly in a n-sized table. Such a table is called a *wavetable*.

As an example consider the following code that uses an array to calculate a sine-wave sampled as 16 points and having a cycle of 2π:

```
1  // a 16-point array, freq = 1 in 2pi
2  ~sig = Array.fill(16, {|i| sin(i/16*2pi)});

4  // posting the table
5  ~sig.do{|i,j| ("["++j++"] = "++i).postln } ;
```

The table is the following:

```
 1  [0]  = 0
 2  [1]  = 0.38268343236509
 3  [2]  = 0.70710678118655
 4  [3]  = 0.92387953251129
 5  [4]  = 1
 6  [5]  = 0.92387953251129
 7  [6]  = 0.70710678118655
 8  [7]  = 0.38268343236509
 9  [8]  = 1.2246467991474e-16
10  [9]  = -0.38268343236509
11  [10] = -0.70710678118655
12  [11] = -0.92387953251129
13  [12] = -1
14  [13] = -0.92387953251129
15  [14] = -0.70710678118655
16  [15] = -0.38268343236509
```

The resulting signal is shown in Figure **8.1**.

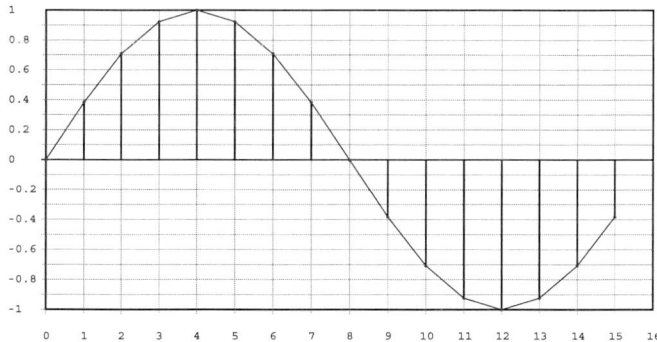

Figure 8.1   Sampled waveform.

If we assume that such a data structure represents a table, a digital oscillator basically performs two operations:

1. reads the sample values from a part of memory that holds them;

2. when it arrives at the final sample (that is 15 in our example) it goes back to the first (0) and starts over again. This last operation is defined as *wrappping around*.

The described synthesis method (*Table Look-Up Synthesis*) is extremely efficient: reading the values from memory is, indeed, several degrees of magnitude faster than calculating the values of a function. Storing information in memory and later retrieving it is, among other things, the very same idea that we discussed when introducing envelopes. The table is nothing but a static model (loaded at initialization): it is then up to the user to decide the amplitude and frequency of the synthesized signal. Regarding the amplitude, the operation should be obvious: we can simply multiply the signal with a scalar so that given the original sample stored in the lookup table of $[-1, 1]$, the output ones will be scaled accordingly (e.g. to a $[-0.5, 0.5]$ range for a multiplication factor of 0.5, Figure **8.2**).

```
1 // a 16-point array, freq = 1 in 2pi, amp = 0.5
2 ~sig = Array.fill(16, {|i| sin(i/16*2pi)*0.5});
```

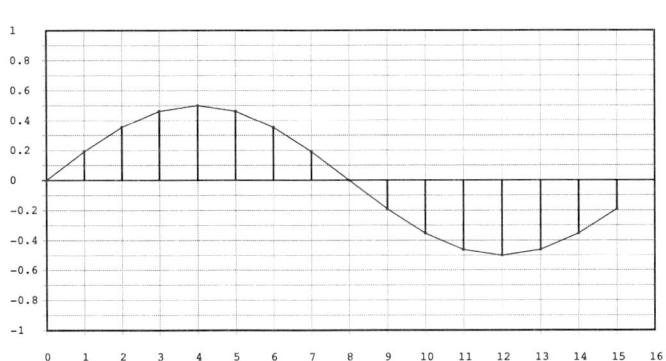

Figure 8.2   Amplitude scaling.

As far as sinusoidal signals are concerned, we can always use the specialized SinOsc UGen which is implemented using table look-up synthesis. The internal table has a size of 8192 samples[2]. However, the use of wavetables is not limited in implementing sinusoidal waves: it can be a more general approach for the synthesis of all sorts of periodic signals.

In SC, there are a number of specialized UGens that can utilize wavetables, the most basic and important of which is Osc—its first argument being a wavetable's buffer ID (the buffer should a have size equal to some power of 2, for reasons of optimization), and the second being the frequency in which the table should be read through. If the table represents a single period from the desired signal, the look-up frequency will actually represent the frequency of the audio signal. The wavetable must be loaded into a buffer that can then be used by Osc. As we saw when discussing the server, a buffer corresponds to a location in temporary memory which the server allocates from RAM. Every buffer, like every bus, corresponds to a numerical address that we use when referring to it.

SC implements a Buffer class that manages this memory. In other words, using Buffer we can ask the server to reserve space from RAM and to load data (and to perform certain operations on them, if needed). Buffer features a number of different allocation/filling methods; among others, there are dedicated methods to read an audio file directly into the buffer as well as methods to fill the space in memory with specific types of signals that are generally useful for sound synthesis.

```
1  (
2  var buf = Buffer.alloc(s, 2.pow(10)) ;
3  // 1 component with normalized amplitude
4  buf.sine1([1]) ;
5  {Osc.ar(buf, 440)}.play ;
6  )

8  (
9  var buf = Buffer.alloc(s, 2.pow(10)) ;
10 // 20 components with random amplitudes
11 buf.sine1( Array.fill(20, { 1.0.rand }) ) ;
12 {Osc.ar(buf, 440, mul: 0.6)}.play
13 )
```

---

[2] Another UGen for sine generation is FSinOsc, but is actually implemented differently, which is why there are certain limitations on its use.

The code above comprises two examples. In the first, we rely on the `alloc` method that asks the server s[3] to construct a buffer of specific size for us. The dimension of the buffer is specified internally as a power of 2 (`2.pow(10)`). Then, by means of sending a `sine1` message to buf, we fill it with a sum of harmonically-related sinusoidal signals. The syntax is comparable to that of `Array`'s `waveFill` method: an array with the amplitudes for each harmonic in the desired signal. In the first example we only load the fundamental of a sine-wave. `Osc`'s first argument is the identifier of the table to scan. Using the frequency argument, the table will be scanned with a frequency of 440 times per second. In the second example, the amplitude array is generated employing `Array`'s `fill` method which, in this particular case, returns an array of 20 pseudo-random numbers within a $[0.0, 1.0]$ range: this stochasticly represents the amplitudes of the first 20 harmonic partials. It is also possible to read a wave-table directly from the hard disc.

In the following example we first generate a signal using the `Signal` class and its `sineFill` method, which generates a sum of sinusoids: it is noted that the dimension of the array (the first argument of the `sineFill` method) is immediately calculated as a power of 2, so that $2^{16} = 65536$. Then, the signal has to be converted into the appropriate format, like that of the `Wavetable` class, so that it may be read by `Osc`: this is done using the `asWavetable` method. The following SynthDef simply bundles `Out` and `Osc` UGens and passes buf as an argument.

Finally, after having stored the signal as an audio file (10–14), it is possible to load it into a buffer: in line 28 we use the `read` method that first allocates a buffer on the server s and then immediately loads the file specified by the path argument. The buffer is assigned to the variable buf. The buffer size (the memory it occupies) is inferred so that it equals the size of the specified file. Thus, the newly generated `Synth` (31) uses the buf buffer.

---

[3] Remember that the global variable s by default represents the audio server.

```
 1  (
 2  var sig ;
 3  var soundFile ;
 4  //--> generating a signal
 5  sig = Signal.sineFill(2.pow(16), [1]) ; // 65536
 6  sig = sig.asWavetable ; // mandatory!
 7  //--> writing the audio signal on a file
 8  soundFile = SoundFile.new ;
 9  soundFile.headerFormat_("AIFF").sampleFormat_("int16").numChannels_(1) ;
10  soundFile.openWrite("/Users/andrea/musica/signalTest.aiff") ;
11  soundFile.writeData(sig) ;
12  soundFile.close ;
13  )

15  (
16  //--> reading synthDef
17  SynthDef("tableOsc",{ arg buf, freq = 440, amp = 0.4 ;
18          Out.ar(0,
19                Osc.ar(buf, freq, mul: amp))
20      }).add ;
21  )

23  (
24  var freq = 440 ;
25  var buf, aSynth;

27  //--> allocating a buffer e immediately filling with file
28  buf = Buffer.read(s, "/Users/andrea/musica/signalTest.aiff") ;

30  //--> reading from buffer
31  aSynth = Synth.new("tableOsc", ["buf", buf]) ;
32  )
```

This oscillator generates samples by means of scanning the wavetable: the waveform stored in the buffer does not have to be sinusoidal. It is, indeed, possible to store in the buffer any kind of waveform. It is possible as well to load a part of any pre-existent signal.

In the following example, the wavetable is filled with pseudo-random values: in line 11 each element of the array (of the Signal type) is assigned to a randomly chosen value within a $[0.0, 2.0] - 1 = [-1.0, 1.0]$ range. The resulting

Signal is then converted to a Wavetable. The rest of the code makes use of the previously used SynthDef (tableOsc). The size of the table (line 11) is determined by exp (line 10), the exponent of 2. Any oscillator that does not immediately stop after the wavetable is over (in which case there would be not point in discussing frequency), by definition produces a periodic signal. Regarding the wavetable, the greater the table and the more complicated its "shape", the more noisy the resulting signal. Conversely, smaller wavetables will produce signals with a simpler time profile (less randomness involved), which would be, therefore, "less dissonant". Try for example to vary exp in the $[1, 20]$ range while visualizing the contents of the buffer using plot. In the lines 23 and 24, we can modulate the frequency according to which the wavetable is scanned; smaller values will make the noisy profile of the latter more evident.

```
1  /*
2      Periodic-aperiodic oscillator:
3      reads from a pseudo-random value table
4  */

6  (
7  var sig, exp ;
8  var soundFile;
9  // generating the table
10 exp = 6 ;          // try in range [1, 20]
11 sig = Signal.fill(2.pow(exp), {2.0.rand-1}).asWavetable ;
12 sig.plot ; // visualizing the table
13 //--> writing the signal on file
14 soundFile = SoundFile.new ;
15 soundFile.headerFormat_("AIFF").sampleFormat_("int16").numChannels_(1) ;
16 soundFile.openWrite("/Users/andrea/musica/signalTest.aiff") ;
17 soundFile.writeData(sig) ;
18 soundFile.close ;
19 )

21 ~buf = Buffer.read(s, "/Users/andrea/musica/signalTest.aiff") ;
22 ~aSynth = Synth.new(\tableOsc , [\buf , ~buf, "amp", 0.1]) ;
23 ~aSynth.set(\freq , 10) ;
24 ~aSynth.set(\freq , 1) ;
```

The oscillator's frequency, thence, indicates how fast/slow the wavetable is read. Specifically, the value passed in tells the oscillator how many times per

second to read through the table. In the following SynthDef we modulate this
rate with the aid of LFNoise0 in a way we have already described before.

```
1  // modulating the freq
2  SynthDef(\tableOsc ,{ arg buf = 0, freq = 440, amp = 0.4 ;
3      Out.ar(0, Osc.ar(buf,
4          LFNoise0.ar(10, mul: 400, add: 400), mul: amp))
5  }).add ;
```

8.1.1 Synthesis by sampling

A conceptually simpler audio synthesis technique is that of sampling. Essentially, a "sample"[4] is an audio signal, typically short in duration, that has been
either directly recorded or extracted from an audio file through *editing*. In any
case, sampling synthesis revolves around the playback of pre-existent audio.

Basic sampling looks a lot like a table-lookup synthesis. As we already discussed, however, the latter primarily concerns the repetition of rather simple
single-period waveforms (albeit it is possible to use any kind of waveform really). Instead, sampling deals with more complex audio excerpts, the duration
of which only depends upon hardware limitations. Regardless of the sample's
origin, once loaded into the memory it may be reproduced whenever needed
through some specialized kind of UGen. Despite this conceptual simplicity,
such an approach is powerful since it enables us to access all sorts of already
existent sounds. These days one can easily find sampled sound of all sorts bundled as libraries, e.g. an entire drum set or the sounds of various orchestral
instruments[5].

---

[4] Note that a "sample" in this context has a different meaning that what we have
hitherto encountered—that is, when discussing sampling rate.

[5] While this is not the place to advertise websites, it is worth mentioning *The
Freesound Project* which features tens of thousands of indexed audio samples
available under a Creative Commons license (http://freesound.iua.upf.edu/).

Although it is possible to playback the samples using an oscillator such as Osc, a specialized UGen for this purpose exists, namely PlayBuf. With the latter we can use buffers of any possible size and not solely those with sizes that correspond to some power of 2.

```
1  (
2  SynthDef(\playBuf , { arg buf, loop = 0 ;
3      Out.ar(0, PlayBuf.ar(1, buf, loop: loop) )
4  }).add ;
5  )
6  (
7  var buf, aSynth ;
8  buf = Buffer.read(s, Platform.resourceDir +/+ "sounds/a11wlk01.wav") ;
9  aSynth = Synth(\playBuf , [\buf , buf, "loop", -1]) ;
10 )
```

In the previous example, the SynthDef simply "wraps" PlayBuf which exposes three of its available arguments, the first two and the last.

- the first argument indicates the number of the channels (typically mono or stereo): if its value is 1, the buffer should also be mono. Note that differences between the indicated and the real number of channels could lead to failure, even if the interpreter will not always report it. Rather than crashing, many buffer reading UGens may just fail silently;
- the second argument specifies the buffer to read from, here buf;
- loop argument indicates the mode of playback and can have either of two values: 0 stands for single-shot mode, and 1 for looping (cyclic playback).

After a buffer is allocated to buf, audio from the hard drive can be read into it. The use of Platform.resourceDir is a cross platform utility method that always points to SC's resources directory. The a11wlk01-44_1.aiff file is included with SC. Even if this file is sampled in a sampling rate of 44.100 Hz (like its name implies), SC will play it back on the server's sampling rate (also

44.100 Hz by default): if those two differ the file will be played back faster or slower than expected.

8.1.2 Resampling and interpolation

When simply reading a buffer with PlayBuf, the result will be an identical signal to the one loaded in the former. In the case of Osc there is a way of specifying the frequency according to which the buffer should be read, meaning that the wavetable is read through in terms of cycles per second. It would be possible to vary the speed according to which we read a buffer with PlayBuf by changing the server's sample rate: but this would be highly inefficient, however, since we cannot modulate the server's sampling rate in real-time. There is, indeed, a more efficient way. Using PlayBuf's rate method we can "tune" a sample in much the same way as we did with wavetables. This process is typically referred to as *resampling*.

Assuming that a simple wavetable has a period $T = 16/44100 = 0.00036281...$ seconds. If looped, the resulting signal's frequency would be $f = 1/T = 2756.25$ Hz. If instead we only read one sample every 2 available ones, the new period would be $T = 16/2 \rightarrow 8/44100 = 0.00018140...$ seconds and, therefore, the frequency of the resulting signal would be $\rightarrow f = 1/T = 5512.5$ Hz, that is two times the original frequency. Reading the table in this fashion essentially "skips" every other value in the table. Figure **8.3** (a) demonstrates a sinusoidal signal with a frequency of $\frac{8}{T}$, sampled at $T = 512$ points. Reading one sample every two results in the signal in Figure **8.3** (b-c): In (b), every cycle of the downsampled signal is only represented by half the number of samples that were present in the original; in (c) we see that within the same duration the number of peaks and valleys of the resulting signal are doubled. That is to say that the downsampled signal has a frequency of $\frac{16}{512}$ and a duration for the cycle is equal to half of the original.

If the reading speed is 2 the frequency is doubled and, as a result, transposed an octave up, while the relationship 4 : 1 indicates a transposition of 4 octaves up. This operation is called *downsampling*. *Upsampling* then, is a similar operation where we make it is possible to lower the frequency of the original sample: for each of the original samples in the table an intermediate one is somehow generated; The resulting frequency will be half of the original one

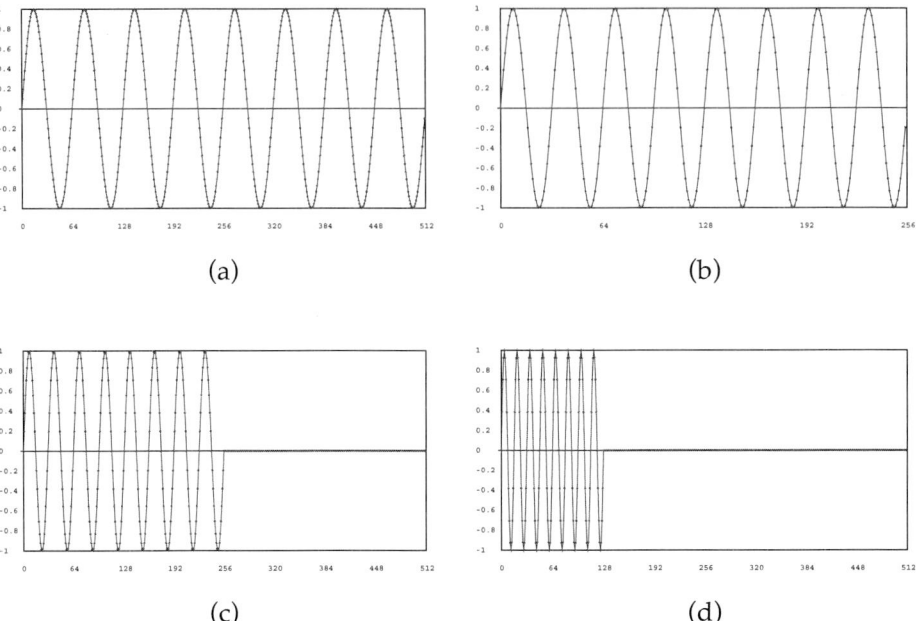

Figure 8.3 Sinusoidal signals with frequency f and period T (a), downsampled version of this signal (reading a sample every two) with $2f$ and $\frac{T}{2}$ (b-c), downsampled version of the signal with $f = 4$ and $\frac{T}{4}$ (d).

which is the musical equivalent of a transposition of one octave lower. In this case, the added samples are not present in the original wavetable but are instead produced with respect to the existent ones. These are typically inferred by means of an interpolation function that takes into account some number of the original samples around the missing sample point that we want to calculate. There are numerous interpolation functions that would be valid in this context: the simpler being a linear interpolation scheme which is the geometrical equivalent of drawing a line between two adjacent points and retrieving the value in the middle. For example, if the first value is 0 and the second 1, then the inferred value would be 0.5. Other interpolation schemes –not to be discussed here– are also possible; usually these are based on the same geometrical principle where the line could also be curved and a value inferred from the contribution of more values.

Interpolation also enables the determination of the value of samples with any fractional relationship between two samples that already exist: that is, we are able not only to retrieve an extra sample every two, three, etc, but also at distances of 0.9, 0.234, between two given samples. This means that we can transpose the original sample in terms of all kinds of interval ratios, including integer and non-integer ones. For instance, the ratio between two frequencies that form a semitone interval (on a guitar fretboard, the distance between two consecutive frets, e.g. between C and $C\sharp$) is $\sqrt[12]{2} \approx 1.06$. In SC we can easily calculate such ratios (remember that in midi notation 60 indicates the middle C) and this way easily represent the semitone:

```
1   61.midicps / 60.midicps // -> 1.0594630943593
```

In order to obtain a semitone transposition it would be necessary to get an extra sample every ≈ 1.06 samples. Such an operation is possible via some interpolation scheme. It has to be noticed, however, that as far as timbre is concerned, doubling the frequency is very different than instrumentally transposing notes; if one is interested in properly emulating the sounds of real instruments, it is more appropriate to sample many different registers of an instrument and limit the amount of transposition that is needed to cover an entire range of notes.

With an oscillator like Osc, the idea is to calculate ahead of time the samples needed for a signal's period, then control its frequency. The frequency is expressed in terms of an absolute value that indicates the number of times the entire wavetable is to be read every second. PlayBuf, on the contrary, is designed to read buffers that contain more complex signals where we do not necessarily know (or care about) the size of the buffer (and prefer not to think in terms of a cyclical period). In this case the frequency is expressed proportionally to the original sound, or in other words, in terms of how fast or slow the sample is played back to provide the opportune adjustment. This relationship can be formalized as $\frac{T_1}{T_2}$, where T_1 is the number of samples a buffer contains and T_2 the number of samples of the resulting signal. The same relationship can be also expressed in terms of frequency with $\frac{f_2}{f_1}$, where f_1 indicates the frequency of the original signal and f_2 that of the resulting one. PlayBuf lets us specify such relationships in terms of its rate argument: if $rate = 1$, the buffer is read according to the server's sampling rate; if $rate = 0.5$, then $T_2 = 2$ and, thence, the buffer is effectively "stretched" to two times its original duration and the resulting signal will have a frequency of f_2—that is an octave down with respect to the original. If rate is negative, the buffer is played back in reverse, from the last sample to the first, and at the speed the argument indicates.

In the following example, the rate argument is given in terms of the output of a Line UGen. The last generates a signal that linearly goes from 1 to -2. This means that the buffer will be initially read at its original $rate = 1$ (given, of course, that the rate according to which the signal has been sampled equals the server's sampling rate) and is then progressively played at a slower speed (this will also result in the sample having greater duration) until $rate = 0$. From this point and on, the playback speed becomes progressively greater, but the buffer is read in reverse. The final value generated from Line is -2, which stands for twice the original playback speed and for reverse playback mode. Note that the synth will be immediately deallocated once Line reaches this value, given that doneAction is set to 2.

```
1  (
2  SynthDef(\playBuf2 , { arg buf = 0;
3      Out.ar(0, PlayBuf.ar(1, buf,
4          rate: Line.kr(1, -2, 100, doneAction:2), loop: 1))
5  }).add ;
6  )

8  (
9  var buf, aSynth ;
10 buf = Buffer.read(s, Platform.resourceDir +/+ "sounds/a11wlk01.wav") ;
11 aSynth = Synth(\playBuf2 , [\buf , buf]) ;
12 )
```

The following example demonstrates the expressive power and efficiency of SC while drawing upon several aspects that we have already discussed. The array source includes 100 stereophonic signals generated by means of Pan2 and PlayBuf. The variable level is an amplitude factor that has been determined empirically. A LFNoise0 UGen controls the rate argument of each PlayBuf, so that it modulates within a $[-2.0, 2.0]$ range—an octave up in both normal and reverse playback modes. Accordingly, all of the 100 PlayBufs read from the very same buffer, each having its playback speed modulated in a pseudo-random way by LFNoise0. The stereo signals are then grouped using flop (line 13) and mixed (line 14-15) so that eventually an array of just two channels is sent to Out.

```
1  (
2  SynthDef(\playBuf3 , { arg buf = 0;
3      var left, right ;
4      var num = 100 ;
5      var level = 10/num ;
6      var source = Array.fill(num, { arg i ;
7          Pan2.ar(
8              in: PlayBuf.ar(1, buf, rate:
9                  LFNoise0.kr(1+i, mul: 2), loop: 1),
10             pos: LFNoise0.kr(1+i),
11             level: level) ;
12      }) ;
13      source = source.flop ;
14      left = Mix.new(source[0]) ;
15      right = Mix.new(source[1]) ;
16      Out.ar(0, [left, right])
17  }).add ;
18  )

20  (
21  var buf, aSynth ;
22  buf = Buffer.read(s, Platform.resourceDir +/+ "sounds/a11wlk01.wav") ;
23  aSynth = Synth(\playBuf3 , [\buf , buf]) ;
24  )
```

8.2 Direct generation

Unlike sampling, the synthesis methods we will present from here on do not rely on pre-existent audio material, but rather revolve around the generation of signals by means of mathematical computations. Emphasis is put, therefore, on how to generate a signal *ex-novo* and not on its subsequent modulation.

8.2.1 Synthesis by fixed waveform

The conceptually simplest method to generate an acoustic signal is with respect to a periodical mathematical function which describes its (*fixed waveform*) in terms of sinusoidal integrals. Even if it is possible to simply calculate the amplitude values for each individual partial, the already described *Table Look-Up* method is more often used: a waveform is stored into a buffer and is read cyclically. It is then possible to generate all kinds of waveforms and not only sinusoidal-based ones: e.g. square, triangle, sawtooth, etc. However, in these last cases a digital oscillator is not necessarily the most optimal solution. For example, certain square or rectangular waveforms can be simply described in terms of a fixed amplitude value and of a particular frequency (or temporal interval) according to which the former alternates. Apart from SinOsc, SC offers a wide variety of periodic signal generators. For example Pulse, that generates square pulses having a variable *duty cycle*—the term indicates the relationship between the positive and negative parts of the waveform's cycle—Saw, that generates sawtooth waveforms, and Impulse, that generates a signal comprised of a single sample repeated at a given frequency. Several of those cases also have dedicated control UGen counterparts (e.g., LFPulse, LFSaw).

An interesting case is noise generators. Noise is a signal where a given sample cannot be predicted in a deterministic way—that is, its amplitude value varies pseudo-randomly with respect to some particular formula: "pseudo", because computers are only capable of deterministic calculations. Thus, noise generators rely on complex formulas that essentially emulate stochastic operations within the confines of computer hardware—such operations are not simple to define. Acoustically speaking, noise is an umbrella-term that accounts for a wide range of diverse phenomena which, nevertheless, have some common sonic features. In the case of white noise, the amplitude values vary in a completely random fashion: the resulting spectrum has energy distributed uniformly in all frequencies. There are other kinds of "colored' noises with different spectral features: brown noise, for instance, is characterized by a constant amplitude attenuation of 6 dB per octave. Those two spectra, obtained by WhiteNoise and BrownNoise UGens, respectively, are demonstrated in Figure **8.4** (a) and (b).

Another interesting noise generator is Dust, which distributes single samples in time in a non uniform way and according to a probabilistic density distribution. In Figure **8.4** (c) we can see such a signal with a density of 100; compare this waveform with that of Impulse, which also produces single samples but in a periodic fashion. The code is the following:

```
1  // white vs pink noise
2  {WhiteNoise.ar(0.5)}.play ;
3  {BrownNoise.ar(0.5)}.play ;
4  // Dust vs Impulse
5  {Dust.ar(100)}.play ;
6  {Impulse.ar(100)}.play ;
```

The output of a pseudo-random number generator can be used as an audio signal on its own, as a source signal that can be further processed (see later, subtractive/additive synthesis) or as a control signal to modulate some synthesis parameter—this way introducing stochastic behavior into deterministic systems.

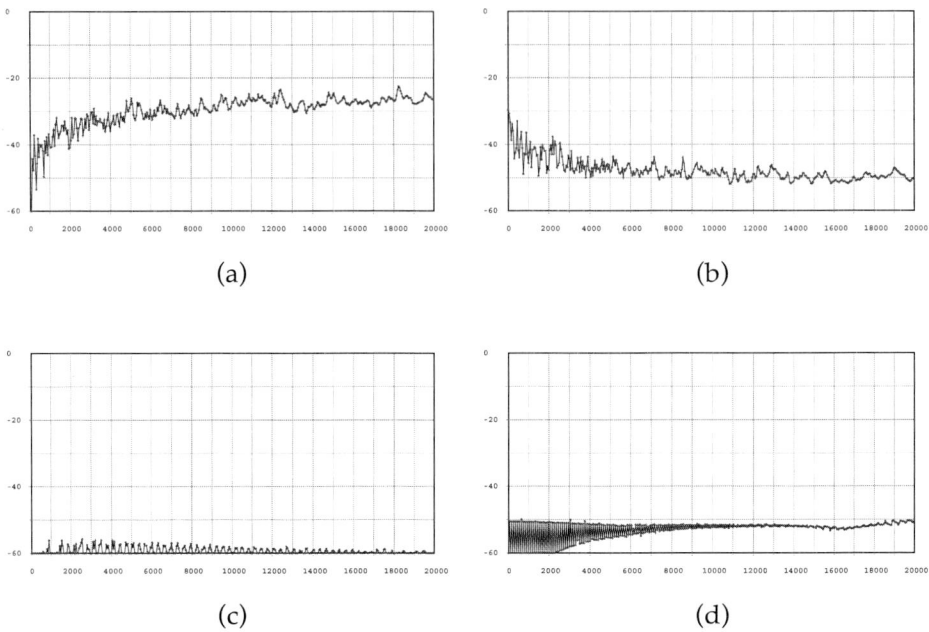

Figure 8.4   Spectra: WhiteNoise (a), BrownNoise (b), Dust (c) e Impulse (d), with density/frequency = 100.

8.2.2   Modulation

We may speak of modulation whenever an oscillator's argument (e.g. amplitude, frequency, phase) is varied according to some other signal. The signal which is modulated is referred to as the *carrier* and the signal that is doing the modulating as the *modulator*. While the simplest oscillator is defined by fixed amplitude and fixed frequency, a modulating version would have one, or both, of these parameters controlled by the output of some other oscillator or dynamic signal. Tremolo and vibrato are two examples where the carrier's amplitude and frequency, respectively, is controlled by another periodical signal. Amplitude and frequency modulation, in their simplest form, are implemented in the same fashion as tremolo and vibrato are. In the case of tremolo and vibrato the modulation takes place in sub-audio ranges—the modulator's frequency typically is lower than 20Hz so that modulation is not perceived as a change of timbre. Faster modulating frequencies will begin to perceptually alter the spectral disposition of the resulting output creating more complex timbres.

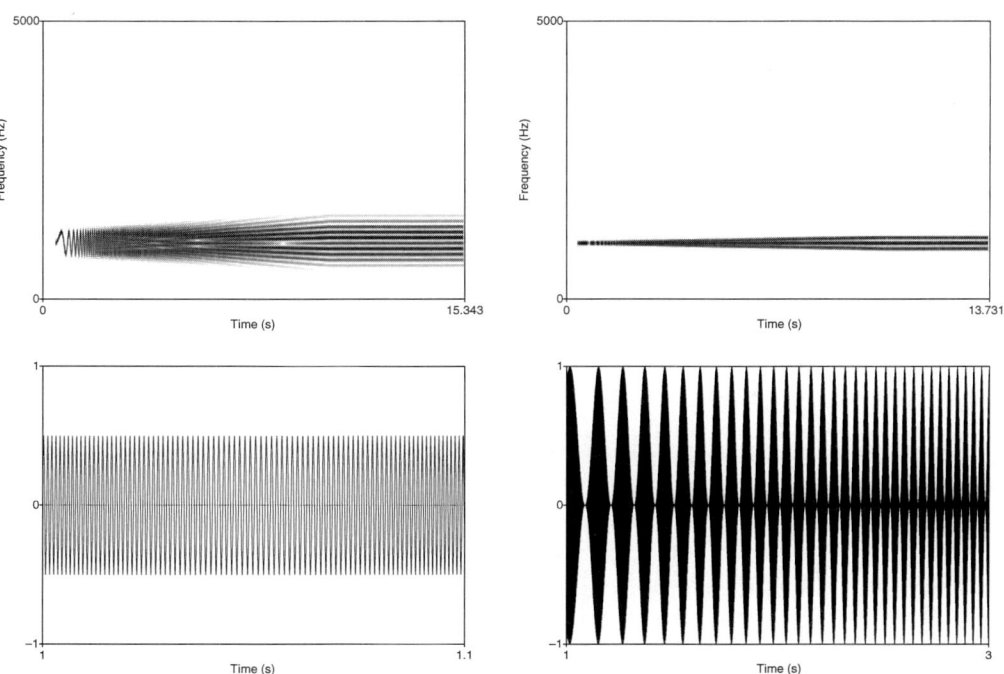

Figure 8.5   Increase of vibrato (left) and tremolo (right) frequency: sonogram (top) and waveform (bottom).

In general, the result of the signal modulation is a new signal where the final sonic features depend upon the frequency and the amplitude of both the carrier and the modulator. Modulation is a widely used technique because it is convenient, predictable and computationally cheap: while additive synthesis requires several oscillators (see later), each with its own control parameters, modulation only requires a couple of oscillators to create signals of equal or greater complexity.

8.2.3   Ring and Amplitude modulation

When the modulating signal (M) controls the amplitude of the carrier (C), we generally consider there to be two different types of amplitude modulation, depending on the characteristics of the modulator. A bipolar signal would cause the carrier's amplitude to vary between a positive and a negative value, while an unipolar one between positive values only. Consider the next example where both unipolar and bipolar approaches are demonstrated for a sine wave with a frequency of 440 Hz—notice the use of the `unipolar` method.

```
1  // bipolar
2  {SinOsc.ar}.plot(minval:-1, maxval:1) ;
3  // unipolar
4  {SinOsc.ar(mul: 0.5, add:0.5)}.plot(minval:-1, maxval:1) ;
5  {SinOsc.ar.unipolar}.plot(minval:-1, maxval:1) ;
```

Audio signals are typically bipolar (having an amplitude typically normalized in the $[-1.0, 1.0]$ range). To transform a bipolar signal to a unipolar one it is sufficient to multiply it with a factor and to add it with an offset (in this example: $[-1.0, 1.0] \rightarrow [-0.5, 0.5] \rightarrow [0.0, 1.0]$).

When the signal that modulates the amplitude is bipolar, we may refer to this as *Ring Modulation* (RM); whereas a unipolar modulator is usually referred to as simply *Amplitude Modulation* (AM). Assuming a carrier and a bipolar modulator with frequencies of C and M, respectively, the resulting signal would create two spectral components (referred to as *side-bands*) of $C - M$ and $C + M$. If $C = 440$ Hz e $M = 110$ Hz, the resulting spectrum would have two components of 330 and 550 Hz. A unipolar modulator would result in a signal where the carrier's original frequency is also present. In the above case, keeping all other parameters the same, a unipolar modulator creates a spectrum with the following partials: $C - M = 330$, $C = 440$, $C + M = 550$. Note that a coefficient with a negative value is to be understood as a coefficient with inverted phase—such a partial would "rival" the positive equivalent; accordingly, a coefficient of -200 Hz would be represented as a having a frequency of 200 Hz but would also cause a coefficient of 200 Hz to attenuate.

The following example demonstrates two possible implementations with otherwise identical results. In the former (3–7), the modulator controls the carrier's `mul` argument. In the latter, the two signals are multiplied together. The `mul` argument indicates a value according to which the signal is to be multiplied,

accordingly the two implementations essentially represent the very same operation and differ only syntactically.

```
1  // better set logarithmic scale in visualization
2  // RM: 2 components
3  { SinOsc.ar(440, mul: SinOsc.ar(110))}.freqscope ;
4  { SinOsc.ar(440)*SinOsc.ar(110) }.freqscope ; // the same

6  // AM: 3 components
7  { SinOsc.ar(440, mul: SinOsc.ar(110, mul: 0.5, add:0.5))}.freqscope ;
8  { SinOsc.ar(440)*SinOsc.ar(110, mul:0.5, add:0.5) }.freqscope ; // the same
```

8.2.4 Ring modulation as a processing technique

AM is typically understood as a technique to build complex spectra from scratch
and by means of simple sinusoidal signals. RM is rather understood as a technique to further process existent signals[6], given its long history in analogue
synthesis[7]. Consider modulating a more complex signal (e.g. of instrumental
origin) with a sinusoidal one: the former would have a spectrum made up of a
series of partials C_n:

$$C = C_1, C_2, C_3..., C_n$$

Given a sinusoidal M, ring modulation $C \times M$ would, therefore, apply to
each component so that the resulting signal would contain all the generated
side-bands—a unipolar M means that the original C would be also present in
the final spectrum:

$$C_1 - M, (C_1), C_1 + M;$$

---

[6] Some references and figures are from Miller Puckette's, *Theory and Techniques of
Electronic Music*.

[7] The name derives from the "ring" configuration of the diodes used to approximate multiplication in analogue synthesizers.

$$C_2 - M, (C_2), C_2 + M,$$
$$\cdots$$
$$C_n - M, (C_n), C_n + M$$

With a smaller M, $C_n - M$ and $C_n + M$ would be closer to C_n: for example, an inharmonic spectrum of $Cn = 100, 140, 350, 470, etc$ and $M = 10$, results in $90, 110, 130, 150, 340, 360, 460, 480, etc - M, etc + M$. Essentially, the spectral envelope remains the same but with two times the original density. Conversely, greater M would cause a spectral expansion. In the previous example, if $M = 2000$ the resulting spectrum would comprise $1900, 2100, 1860, 2140, 1650, 2350, 1730, 247$ $M, \ldots + M$—conisder Figure **8.6**.

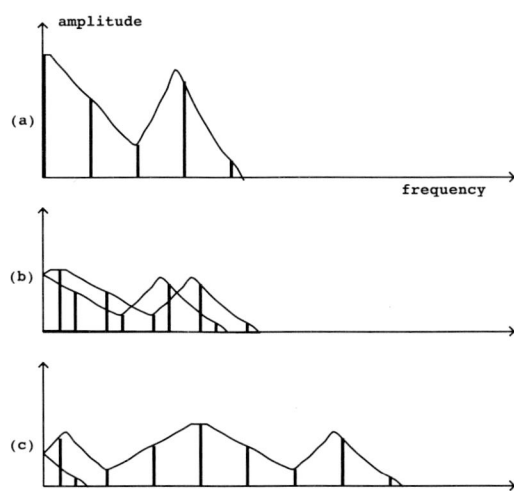

Figure 8.6 M as a modulator of the spectral envelope: contraction and expansion (from Puckette *Theory cit.*).

In the following example an audio file is ring-modulated with a sine wave of a frequency that is exponentially incremented from 1 to 10000 Hz within a period of 30 seconds. Notice the initial effect of "doubling" which eventually leads to the emergence of spectral component around the frequencies of the sine wave.

```
1  b = Buffer.read(s, Platform.resourceDir +/+ "sounds/a11wlk01.wav") ;

3  (
4  { PlayBuf.ar(1, b, loop: 1)
5      *
6      SinOsc.ar(XLine.kr(1, 10000, 30))
7  }.freqscope
8  )
```

The same approach is valid when M is a complex signal: this is often the case in analogue RM, when two complex signals modulate each other[8]. When both C and M are complex signals, side-bands are reciprocally generated for all component combinations. If C and M have i and k spectral components, respectively, then $C \times M$ would have $i \times k$ components—a signal with great complexity.

The following example employs the same audio file we have used hithero. The first modulator is a copy of this file having its playback speed modulated exponentially in the $[0.1 - 10.0]$ range (you can listen to this alone by evaluating the first block only). It can be seen that the resulting signal (lines 8–14) has a very complex spectral envelope—traces of the original may be still perceived, nonetheless.

---

[8] For example, this happens in several important works by K.H.Stockhausen: *Kontakte, Hymnen, Mikrophonie, Prozession.*

```
1  ( // the modulating signal
2  { PlayBuf.ar (1, b,
3      rate: XLine.kr (0.1, 10.0, 30, doneAction:0),
4      loop: 1) }.play ;
5  )

7  (
8  {
9      PlayBuf.ar (1, b, loop: 1)
10     *
11     PlayBuf.ar (1, b,
12     rate: XLine.kr (0.1, 10.0, 30, doneAction:0),
13     loop: 1)
14     *5
15 }.freqscope
16 )
```

In the following example, the same audio file is multiplied with a quadratic waveform having a frequency that increments from 0.5 (so that $T = 2$) to 100 Hz in 30 seconds. Note that when the frequency is in the subaudio range (where $freq <\approx 20$), the modulator largely operates as a percussive envelope.

```
1  {      PlayBuf.ar (1, b, loop: 1)
2      *
3      Pulse.ar (XLine.kr (0.5, 100, 30, doneAction:2))
4  }.freqscope
```

Another interesting case occurs when a complex harmonic signal with a fundamental frequency C is used as a carrier and a sinusoidal signal with a frequency of $M = \frac{C}{2}$ as a modulator. The resulting spectrum would be harmonic with a frequency of $M = \frac{C}{2}$ and would only comprise odd harmonics. This, essentially, transposes the carrier down an octave.

For example, if $C_{1-n} = 100, 200, 300, ...n$ and $M = 50$, the resulting signal would consist of the following components:

$$100 \pm 50, 200 \pm 50, 300 \pm 50$$
$$=$$
$$50, 150, 150, 250, 250, 350, 350$$

Which are the odd harmonics of $\frac{c}{2}$; indeed, the ratios between the resulting partials and the original frequency are $1, 3, 5, 7, ...$ In order to retrieve the even harmonics, simply add the original to this signal.

This way we get a simple digital implementation of an analogue effect typically used in pop/rock music, the *octaver*.

```
1  //RM octaver
2  (
3  SynthDef.new(\RmOctaver , {
4      var in, an, freq ;
5      in = SoundIn.ar(0) ;       // audio from mic
6      an = Pitch.kr(in).poll ; //  analysis signal
7      freq = an[0] ;      // the retrieved fundamental freq
8      Out.ar(0, SinOsc.ar(freq: freq*0.5)*in+in);
9      // RM freq/2 + source
10 }).add ;
11 )

13 Synth.new(\RmOctaver ) ;
```

The RmOctaver SynthDef processes the soundcard's input (typically connected to some microphone) via the SoundIn UGen—remember that the first argument of the latter stands for the bus' index and that the index 0 always stands for the first available input. Pitch is a *pitch-tracking* analysis UGen that attempts to detect the fundamental frequency of the signal in its input. Pitch outputs an array of two signals: the first is a value for the detected pitch in Hz while the second indicates whether the algorithm is confident in its analysis. The algorithm's maximum amount of certainty in its analysis results in a value of 1, and if not, this signal will have a value of 0[9]. The frequency $\frac{f}{2}$ is then used to control the oscillator which is ring-modulated with the in signal; the latter is, then, added to the result.

---

[9] Note that analyses UGens typically implement a *kr method only. In fact, to analyze a signal it is necessary to take into account several samples from the input; there is no such thing as the frequency of a single sample.

In a similar fashion, when $M = n \times C$ the carrier's higher-end partials are affected. If $C = 100, 200, 300, n \times 100$ and $M = 200(n \times 2)$, the resulting spectrum would be: $100 - 200 = 100, 100 + 200 = 300, 200 - 200 = 0, 200 + 200 = 400, \dots$

8.2.5 Frequency modulation

In the case of frequency modulation (FM), it is the carrier's frequency which is controlled by the modulator. Considering two oscillators, to implement FM we typically add the output of the modulator to the carrier's frequency C. In this way C's frequency deviates towards greater (when the output of the modulator is positive) and smaller (when the modulator's output is negative) values. The *peak frequency deviation* term, or simply deviation D is often referred to as the index of the carrier's frequency maximum traversal in Hz. A fundamental difference between FM with RM or AM is that with FM, the process produces a theoretically infinite number of side-bands for each $C \pm n \times M$, though in practice the amplitude of these sidebands is strictly controlled based on the index of modulation. If, e.g. $C = 220$ and $M = 110$, then for $n = 1 \rightarrow 330$ and 110; for $n = 2 \rightarrow 440$ and 0; for $n = 3 \rightarrow 550$ and 110 (-110 stands for inverse phase), etc. FM then opens up the way for generating signals of arbitrary complexity using just a couple of oscillators. It is precisely for this reason that FM became the very first audio synthesis technique to be met with commercial success, thanks to a series of Yamaha synthesizers (in particular the famous DX7). FM has been used from the beginning of the 20th century in telecommunications ("radio frequency modulation" is often abbreviated as FM). In the latter 1960s, however, John Chowning—still a student at Stanford University— had been experimenting with very fast vibrati. He eventually implemented a digital version of FM and mathematically formalized it (1973). *Turenas* (1972) is the first piece to be written using FM techniques extensively (it should be noted that the first-ever attempt is 1966's *Sabelithe*, which was finally finished in 1988). Employing just a couple of oscillators, Chowning achieved timbres that would have taken 50 or more oscillators to realize with additive synthesis. Yamaha bought this patent (which came to be the most profitable in the history of Stanford University) and produced a number of synthesizers that revolved around these FM techniques.

FM is characterized by its ample potential for the generation of complex sounds. However, from the theoretically infinite side-bands, just a few are significant: such a number can be determined through the modulation index I.

The modulation index is defined in terms of deviation and the modulator's frequency: $I = \frac{D}{M}$. Following, D can be said to represent the "depth" of the modulation. The value given from $I + 1$ is considered as an approximation of how many "significant" side-bands are present in the resulting FM's output spectrum. The utility of the formula lies in that, if D and M are constants, I, then, constitutes a measure of the output signal's complexity. It follows that the modulator's amplitude is given by $D = I \times M$. If $I = 0$, the deviation is zero and the carrier is not modulated at all. To increment the modulation index is to increase the frequency deviation and, following, to add complexity to the resulting spectrum.

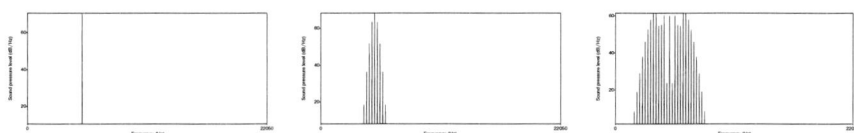

Figure 8.7 $I = 1, 3, 7$.

To summarize: In FM the nature of the generated spectrum (that is, the positioning of the generated side-bands) is determined by the relationship between the carrier and the modulator, while the richness of the spectrum (the number of the generated side-bands) can be thought of in relation to the amplitude of the modulator.

The following two examples are meant to be visualized, rather than listened to. Note that linear spectral plotting would better account for the partials' symmetrical expansion caused by controlling the modulator's frequency with respect to that of the carrier (line 1). In the first example a 5000 Hz carrier is gradually modulated with M moving from 10 to 1000 Hz.

```
1  s.freqscope ; // better set linear visualization

3  (
4  { SinOsc.ar( 5000      // carrier C
5      + SinOsc.ar(XLine.kr(10, 1000, 60, doneAction:2), mul: 1000),
6    mul: 0.5
7      )}.play
8  )
```

The following example pinpoints the modulator's amplitude which draws energy from the carrier's fundamental and distributes it to the various side-bands. Note that the spectral envelope follows a rather symmetrical pattern centered at the carrier, even if the former is not always lance-shaped. The exact shape of the spectral envelope can be predicted through a set of functions referred to as "Bessel functions".

```
1  (
2  { SinOsc.ar(
3        10000      // carrier C
4        + SinOsc.ar(500, mul: XLine.kr(1, 20000, 60, doneAction:2)),
5     mul: 0.5
6        )}.play
7  )
```

The last example is rather minimal as it only employs two oscillators. Herein, using the mouse we can interactively explore the effect of FM and to immediately appreciate the possible complexity of the resulting signal.

```
1  (
2  { SinOsc.ar(
3        2000      // carrier C
4        + SinOsc.ar(    // modulator M
5            freq: MouseX.kr(0, 1200), // freq for M
6            mul: MouseY.kr(0, 20000)    // amp for M
7            ),
8     mul: 0.5
9        )}.freqscope
10 )
```

8.2.6 C:M ratio

As far as the spectral characteristics of FM's output, in addition to the modulator's amplitude, another important factor to take into account is the ratio between C and M. This factor is of particular importance both in the case of FM and complex RM where, unlike AM or simple RM (when both signals are sinusoids), the generated spectra includes numerous components.

The relationship between the frequencies of the two signals is usually referred to as *C:M ratio*, hence on *cmr*. Since the resulting spectrum contains frequencies that are the sums and differences of C and M, an integer *cmr* would create a harmonic spectrum that includes multiples of the component's greatest common divisor. For $C = 5000$ Hz and $M = 2500$ Hz ($cmr = 2 : 1$), the resulting spectrum in AM would comprise 2500, 5000 and 7500 Hz: that is, a harmonic spectrum formulated from a fundamental (2500) and its first two harmonic partials. Two interesting cases in both RM and FM emerge when $cmr = 1 : 2$ and $cmr = 1 : 1$. In the first case only the odd harmonics will be present: with e.g. $C = 1000$ and $M = 2000$, RM results in $(-)1000, 3000$, and FM in $(-)3000, 5000, (-)5000, 7000$, etc. In the second case, all the harmonics are present in the output signal: for $C = M = 1000$, RM results in $0, 2000$, and FM in $0, (-)1000, 3000, (-)2000, 4000$ etc.

If the ratio has a denominator of 1 (such as in the preceding example), the resulting frequencies would be multiples of the modulator's frequency that, as a result, becomes a new fundamental. If the denominator is greater than 1, then the greatest common divider of C and M becomes the fundamental of the generated signal: for $C = 3000$ and $M = 2000$ ($cmr = 3 : 2$) the new fundamental would be 1000 Hz and the spectrum would comprise 1000 Hz $(3000-2000)$, to which we should add (in AM) 3000 Hz (C) and 5000 Hz $(3000+2000)$. In this case it is also possible that the fundamental may be "lost". For example, if $C = 5000$ and $M = 2000$, (that is $cmr = 5 : 2$) the new fundamental is always 1000 Hz (the GCD of 5000 and 2000), yet the spectrum comprises 3000 Hz $(5000 - 2000)$, 5000 (AM), and 7000 $(5000 + 2000)$. The fundamental of 1000 Hz is considered "missing" (also "suppressed" or "phantom"), because even if it is not physically present, it is reconstituted by our ears given that its III, V and VII harmonics are present. This is due a complex psychoacoustic phenomenon according to which our ears reconstruct the missing fundamental once a sufficient number of higher harmonics imply its presence.

The latter is also true for FM, as well as for RM and simple AM where C and M are sinusoidal signals, and where the generated sidebands are only 2 or 3. In the previous example, with $C = 5000$ Hz and $M = 2000$ Hz, AM would produce the following frequencies: 3000, 5000, 7000 Hz, i.e. the 3rd , the 5th, and the 7th harmonic of 1000 Hz. The *cmr* is, therefore, an indicator of how "harmonic" the resulting spectrum is: the simpler the fraction (that is, the lower the product $C \times M$), the denser the resulting harmonics. An almost integer *cmr* (e.g. 2.001 : 1) result in spectra with some inharmonic elements that, nevertheless, sound even more harmonic and "natural"—which is precisely because acoustic instruments are typically characterized by the presence of some inharmonic elements.

In general, *cmr* of $N : 1$ and $1 : N$, will generate the same spectrum, given that the rest of the modulation parameters are identical. The number of partials can be, then, calculated from *cmr*'s components. E.g. if *cmr* is 2 : 3, then $|2 \pm 3 \times n| = 1, 2, 4, 5, 8, 11...$. If $C > 1$, then the resulting spectrum will feature inharmonic elements (or a "lost" fundamental). E.g. If $cmr = 2 : 5$, the resulting spectrum would be $2, 3, 7, 8, 12, 13, ...$, which apparently lacks the fundamental (i.e. 1) as well as many other harmonics. To boot, the spectrum is rather sharp. Consider $cmr = 5 : 7$, which results in a particularly inharmonic spectrum: $2, 5, 9, 12, 16, 19, 23, 26, ...$

The following SynthDef lets us control FM employing, other than the carrier's frequency freq (C), parameters that derive from *cmr*: these are c, m, a. The first two indicate the numerator and denominator of *cmr*, and the third the modulator's amplitude.

```
1  (
2  SynthDef(\cm , { arg f = 440, c = 1, m = 1, a = 100, amp = 0.5 ;
3      Out.ar(0,
4      SinOsc.ar(
5      f      // base freq for carrier C
6      + SinOsc.ar(    // modulator M
7          freq: f * m / c, // freq for M, calculated from cmr
8          mul: a    // amplitude for M
9          ),
10    mul: amp) // amplitude for C
11     )
12 }).add ;
13 )
```

The graphic interface in the following example lets us control f, c, m, a with number boxes. The code produces values for just C and M, internally. Even if a procedural kind of syntax is employed (Array.fill), the code is rather long since each graphic element has to be associated with a synth's parameter.

```
1  var cmsynth = Synth("cm") ;
2  var freq = 2000 ; // f C: 0-2000 Hz
3  var num = 30 ; // ratio for c:m
4  var w = Window("C:M player", Rect(100, 100, 220, 420)).front ;
5  var sl = Array.fill(4, {|i| Slider(w, Rect(i*50+10, 10, 50, 350))}) ;
6  var nb = Array.fill(4, {|i| NumberBox(w, Rect(i*50+10, 360, 40, 20))}) ;
7  ["freq C", "C", "M", "amp M"].do{|i,j|
8      StaticText(w, Rect(j*50+10, 390, 40, 20)).string_(i).align_(\center )
9  } ;

11 sl[0].action = { arg sl ; // base freq
12     var val = sl.value*freq ;
13     cmsynth.set("f", val) ; nb[0].value = val ;
14     } ;
15 nb[0].action = { arg nb ;
16     var val = nb.value ; // 0-1000 Hz
17     cmsynth.set("f", val) ;    sl[0].value = val/freq ;
18     } ;

20 sl[1].action = { arg sl ; // numerator C
21     var val = (sl.value*(num-1)).asInteger+1 ;
22     cmsynth.set("c", val) ;    nb[1].value = val ;
23     } ;
24 nb[1].action = { arg nb ;
25     var val = nb.value.asInteger ;
26     cmsynth.set("c", val) ; sl[1].value = val/num ;
27     } ;

29 sl[2].action = { arg sl ; // denominator M
30     var val = (sl.value*(num-1)).asInteger+1 ;
31     cmsynth.set("m", val) ;    nb[2].value = val ;
32     } ;
33 nb[2].action = { arg nb ;
34     var val = nb.value.asInteger ;
35     cmsynth.set("m", val) ;    sl[2].value = val/num ;
36     } ;

38 sl[3].action = { arg sl ; // amplitude M
39     var val = sl.value*10000 ;
40     cmsynth.set("a", val) ;    nb[3].value = val ;
41     } ;
42 nb[3].action = { arg nb ;
43     var val = nb.value ;
44     cmsynth.set("a", val) ;    sl[3].value = val/10000 ;
45     } ;
```

8.2.7 Waveshaping

The technique of *waveshaping*, also referred to as non-linear distortion, reflects aspects of both wavetable synthesis and modulation.

If a signal is multiplied by a constant k, the structure of its waveform remains unchanged; its amplitude, instead, increases or decreases proportionally to k. Waveshaping is a linear operation, according to which the input waveform (most typically, but not necessarily, a sine wave) is expected to be distorted. Such distortions in the waveform correspond to the addition of harmonic elements to the signal's spectrum. For example, in the case of clipping, a contingency of waveform synthesis which is very famous in pop/rock music contexts, the input waveform is "cut" above a certain threshold to introduce a "squaring" effect in the signal.

In analogue circuits, clipping is the result of failing components which can no longer result in signals that are proportional to their input. In the digital domain, waveshaping occurs with respect to a wavetable that associates the values of the input samples to output ones.

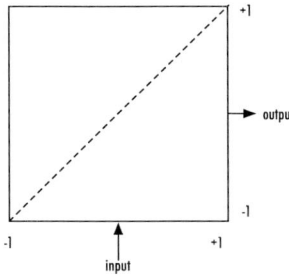

Figure 8.8   Waveshaping: a transfer function stored to a wavetable.

Such a wavetable associates each possible input value to an output one, as shown in Figure **8.8**, and is referred to as "transfer function". If the function is a straight line with a 45° angle against the horizontal axis, each value in the output would be unchanged in the output. But any deviation in the shape of such a line generates distortion of some sort. A possible implementation is shown in the next example.

```
1  // waveshaping
2  t = FloatArray.fill(512, { |i| i.linlin(0.0, 512.0, -1.0, 1.0) });
3  t.plot ;
4  b = Buffer.sendCollection(Server.local, t)

6  {
7      var sig = SinOsc.ar(100) ;
8      Out.ar(0, BufRd.ar(1, bufnum: b,
9          phase: sig.linlin(-1.0, 1.0, 0, BufFrames.ir(b)-1)
10     ))
11 }.scope
```

The array t comprises 512 values scaled within the $[-1, 1]$ range. As shown in the diagram, t is a ramp which can be thought of as a nominal transfer function. Buffer has several methods that we can use to directly send to the server such transfer functions that we have calculated client-side—e.g. using sendCollection we can load t to the buffer b. The following SynthDef assumes a 100 Hz sine-wave as its source. BufRd generates an output that corresponds to the value present in the given buffer (here b) and at the index specified by phase—the latter is modulated by the input signal. In this case, the sine wave controls the buffer's index so that BufRd's output would be an accelerated bi-directional reading of b, as long as the sine wave is properly scaled to within the table's bounds. Since the wavetable consists of 512 point only, the oscillator's output (originally in the [-1,1] range) has to be scaled to a range of [0,511]. Here we rely on the utility UGen BufFrames to calculate this number: the latter returns the number of frames in the buffer; since buffers are indexed from 0, however, we need to subtract 1 from this value to get the actual index of the last frame. Note that the rate of the BufFrames is ir, which stands for "instrument rate"—meaning, output is only updated once, when the synth is allocated.

Under this premise, the next few examples explore ways to generate algorithmic content for the array and demonstrate that the effect of waveshaping is sensitive to the the input's dynamics. The first wavetable squeezes all values below 0.5 and expands all those above it. Then, input values that exceed 1 will be clipped (which is preferable in this context). The normalize method helps keep the input signal within the expected ranges. The overall distortion depends on the input's amplitude (controlled here via the Mouse).

```
1  // 1
2  (
3  t = FloatArray.fill(512, { |i|
4      v = i.linlin(0.0, 512.0, -1.0, 1.0) ;
5      if (abs(v) < 0.5){v*0.5} { v*1.25}
6  }).normalize(-1.0, 1.0);
7  t.plot ;
8  b = Buffer.sendCollection(Server.local, t)
9  )

11  (
12  {
13      var sig = SinOsc.ar(100, mul:MouseX.kr(0,1)) ;
14      Out.ar(0, BufRd.ar(1, bufnum: b,
15          phase: sig.linlin(-1.0, 1.0, 0, BufFrames.ir(b)-1)
16      ))
17  }.scope
18  )
```

The second example employs a very complex table (built empirically) that produces a rich spectrum that is, again, sensitive to changes in dynamics. Note that, at low amplitudes, the wavetable produces signals that are not balanced around zero—that is, signals with "DC offsets". DC offset is a term originating from analog synthesis where it stands for the presence of a direct current (DC) in the signal (by the way, SuperCollider does contain a UGen for removing DC offset, LeakDC).

```
1  // 2
2  (
3  t = FloatArray.fill(512, { |i|
4      v = i.linlin(0.0, 512.0, -1.0, 1.0) ;
5      v.round(0.125+(v*1.4*(i%4)))
6  }).normalize(-1.0, 1.0);
7  t.plot ;
8  b = Buffer.sendCollection(Server.local, t)
9  )

11 (
12 {
13     var sig = SinOsc.ar(100, mul:MouseX.kr(0,1)) ;
14     Out.ar(0, LeakDC.ar(
15         BufRd.ar(1,
16             bufnum: b,
17             phase: sig.linlin(-1.0, 1.0, 0, BufFrames.ir(b)-1))
18     ))
19 }.scope
20 )
```

The Figure shows the two spectra generated when the SinOsc's amplitude is 0.0001 and 1.0, respectively. The two signals have been normalized also so that the differences in their spectral structure, rather than their amplitude, are apparent.

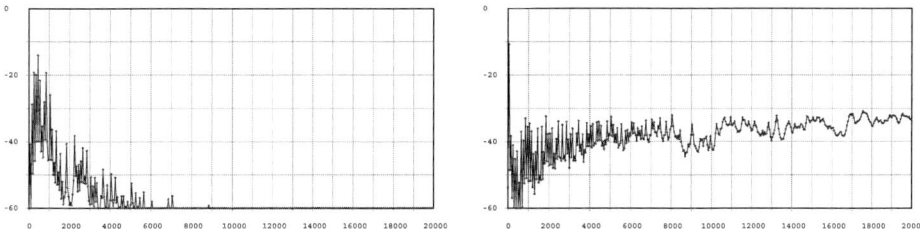

Figure 8.9 Spectra when SinOsc's amplitude is 0.0001 and 1.0.

In reality, the standard way to implement waveshaping in SC is via a dedicated UGen–Shaper– as demonstrated in the next example. The table is built via the Signal class. Note that an instance of Signal must have a size equal to some power of 2 plus 1 (513 in this case) in order to be used as a transfer function with

Shaper. A buffer of double this size (without counting the extra sample in the Signal, that is 1024 in our case) is then allocated. The buffer must be double the size of the original data because Shaper expects buffers with data represented in a special "wavetable" format that optimizes the performance of the UGen. Accordingly, data is converted to a wavetable using the WavetableNoWrap method. Shaper's parameters are rather straightforward: the buffer with the transfer function and the input signal.

```
1  // dimension: power of 2 + 1
2  t = Signal.fill(513, { |i|
3      var v = i.linlin(0.0, 512.0, -1.0, 1.0);
4      if (abs(v) < 0.5){v*0.5} { v*1.25}
5  }).normalize(-1.0,1.0);
6  t.plot ;

8  // double buffer
9  b = Buffer.alloc(s, 1024, 1);
10 b.sendCollection(t.asWavetableNoWrap);

12 { Shaper.ar(b, SinOsc.ar(440, 0, MouseX.kr(0,1)))*0.75 }.scope ;
```

Shaper is particularly useful in order to easily implement a series of specific transfer functions, the so-called Chebysev polynomials, that allow to precisely calculate their effect on some sinusoid signal. Waveshaping typically produces rich spectra which are, nonetheless, hard to understand or deal with theoretically. In this respect, waveshaping is similar to modulation. In the following example a buffer is filled with the appropriate polynomials that will result in the first 20 harmonics of an input sine-wave—each having a pseudo-random amplitude in this case.

```
1  b = Buffer.alloc(s, 1024, 1);
2  b.cheby(Array.fill(20, {1.0.rand}));

4  { Shaper.ar(b, SinOsc.ar(440, 0, 0.4)) }.scope;
```

In conclusion, while with modulation techniques we can easily and efficiently (computationally speaking) generate signals of great complexity (think of e.g. the spectral complexity we can achieve with just two sinusoidal signals), to control the result is not intuitive. More, modulation techniques are rather inappropriate for analysis applications: it is extremely difficult to represent existent audio material in terms of modulation.

8.3 Spectral modelling

Spectral modeling techniques are techniques that generates spectra typically expressed in terms of what spectral components need to be present and with what strength. Accordingly, all of the following techniques foreground the frequency, rather than the time, domain.

8.3.1 Additive synthesis

According to the Fourier theorem, every periodic signal, however complex it may be, can be represented in terms of a sum of simple sinusoidal components. Additive synthesis takes its basis around this simple fact. In additive synthesis, we add a series of sine waves, each having a dedicated amplitude envelope, to synthesize complex signals.

Figure **8.10** illustrates a bank of oscillators that operate in parallel. Their n outputs are added together to result in a complex and rich spectrum. In classic additive synthesis, we use "tuned" oscillators, the frequencies of which ($f2...fn$) follow the harmonic series of some fundamental ($f1$).

Additive synthesis is one of the oldest sound synthesis methods, due to fact that it is fairly simple to implement it. In audio synthesis, when we speak theoretically about "adding", we typically refer to ordinary "mixing". In SC, Mix is a specialized UGen for such an operation; given an array of signals in its input, it will return a new monophonic signal comprised of the sum of all the elements in this array.

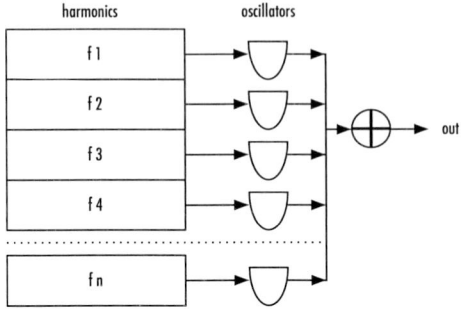

Figure 8.10 Bank of oscillators.

The following example, albeit minimal, is instructive and demonstrates the expressive power of SC. As seen in line 3 we introduce a Mix object with an array of 20 signals (generated by a SinOsc) passed as an argument. Notice how counter i is multiplied with a fundamental frequency (200 Hz) to generate its higher harmonics—of course, since i's initial value is 0, we have to add 1 to achieve the desired results. Mixing is a linear operation which means that if n signals normalized in the $[-1, 1]$ range are mixed, the result will be a signal oscillating in the $n \times [-1.0, 1.0] = [-n, n]$ range: this would result in an unacceptable level of distortion due to clipping our output. In our example, we set mul: 1/20 to scale all signals accordingly so the final mix is always normalized.

```
1  { Mix.new      // mix
2      ( Array.fill(20, { arg i ; // 20 harmonics
3          SinOsc.ar(200*(i+1), mul: 1/20) }))
4  }.scope ;
```

In line with the algorithmic orientation of SC, a specialized fill method for Mix does exist. The following example is acoustically identical to the previous one.

```
1  {
2  // the same
3  Mix.fill(20, { arg i ;
4      SinOsc.ar(200*(i+1), mul: 1/20)})
5  }.scope ;
```

So far we have only used "synced" sine waves, that is, their cycles start at the same time. This causes peaks in the final output. Generally speaking, phase differences in stationary signals are inaudible—the audible system focus on spectral components. But of interest here is the frequency relationships between the various spectral components. However, it is best to avoid having all oscillators in phase (which with enough components, will create an impulse). In the following example, each oscillator operates at a random phase with respect to 2pi.rand to avoid this phenomenon.

```
1  // avoiding phase sync
2  { Mix.fill(20, { arg i ;
3      SinOsc.ar(200*(i+1), 2pi.rand, mul: 1/20)})
4  }.scope ;
```

Varying the amplitude of the harmonic components affects their relative contribution to the synthesized spectrum. Below, a stereo signal is generated as the sum of 40 sinusoids that begin at 50 Hz. The amplitudes of the left-channel components are modulated differently that those of the right-channel ones. The former vary randomly at a rate of 1 per second, while the latter with respect to a sinusoidal oscillator in which the frequency randomly changes within the $[0.3, 0.5]$ range. Then, with the appropriate mul and add values, the signal becomes unipolar. Note that in order to normalize the output, we divide by 20 rather than by 40. Given the amplitude variations of the components, the output is no longer characterized by great peaks and as such we have empirically estimated that a value of 20 would be sufficient to avoid clipping.

```
1  // stereo spectral motion
2  { Mix.fill(40, { arg i ;
3      var right = LFNoise1.kr(1, 1/20)   ;
4      var left = SinOsc.kr(rrand(0.3, 0.5), 2pi.rand, mul: 0.5, add: 0.5) ;
5      SinOsc.ar(50*(i+1), [2pi.rand, 2pi.rand], mul: [left/20, right]) })
6  }.scope ;
```

In the following example, the amplitude of each component is scaled proportionally to the reciprocal of harmonic indexes: the greater the index, the lower the amplitude, following the typical behavior of musical instruments. The array arr is made up of a linear sequence of integers from 1 to 20, the order of which is, subsequently, reversed and normalized so that the sum of all components eventually equals 1. Given that the sum of all elements in arr would be 1, we can safely use the former to set the amplitudes of our oscillators and avoid clipping. The oscillator's frequency is expressed in midi notation which is then converted to Hz through midicps.

```
1  {
2  var arr = Array.series(20, 1).reverse.normalizeSum ;
3  Mix.new     // mix
4      ( Array.fill(20, { arg i ; // 20 partials
5          SinOsc.ar(60.midicps*(i+1), 2pi.rand, mul: arr[i])}))
6  }.scope ;
```

With additive synthesis we can procedurally generate all sorts of harmonic waveforms. For instance, the square wave can be defined as an infinite sum of odd harmonics whose amplitude is proportional to the inverse of the harmonic number, following this equation: $f_1 \times 1, f_3 \times 1/3, f_5 \times 1/5, f_7 \times 1/7...$

As shown in Figure **8.11**, the more harmonics the better the approximation of the square wave. Theoretically speaking, it takes infinite sinusoidal components to generate a square wave; on this respect it is preferable to generate the latter by other means or using the dedicated generators. A sawtooth wave can be also generated in a similar fashion, with the exception that all harmonics, and not just the odd ones, should be present. As seen in the figure, the steepness of the waveforms' edges is proportional to the number harmonics that are

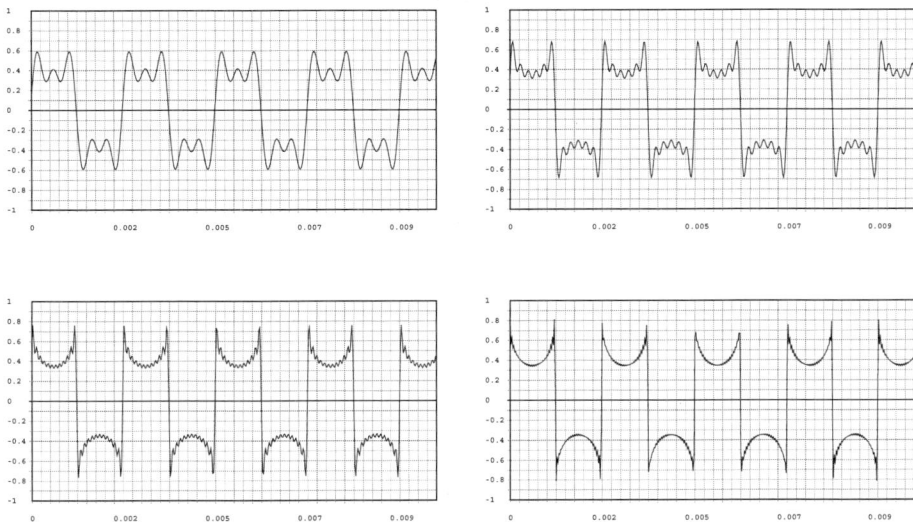

Figure 8.11 Square wave: first 3, 6, 12, 24 harmonics.

present. Generally speaking, more rounded waveforms suggest the presence of a few harmonics only and, conversely, steep ones suggest the presence of many. The following code demonstrates the difference between sawtooth and square waves. The environment variable ~numArm and ~baseFreq allow to experiment with various values.

```
1  ~numArm = 10 ; ~baseFreq = 69.midicps ;

3  // square wave
4  {Mix.fill(~numArm, {|i|
5      SinOsc.ar(~baseFreq*(i*2+1), mul:1/( i+1))
6  })}.play ;

8  // sawtooth wave
9  {Mix.fill(~numArm, {|i|
10     SinOsc.ar(~baseFreq*(i+1), mul:1/( i+1))
11 })}.play ;
```

It is not accidental that we first introduced `Mix` when discussing additive synthesis. In fact, the additive paradigm can be generalized in terms of micro-mixing/editing: the idea is to generate complex signals on the account of simpler ones that are mixed together. In this case, a sum of partials stands for arbitrary spectra and not necessarily for harmonic ones. The latter may as well include inharmonic components and/or noises. Then, the sinusoidal elements of such a spectrum can be understood in terms of their harmonic `ratios`, much like the way cmr was used to understand the relationship between carrier and modulator in the FM paradigm.

Accordingly, harmonic signals are characterized by an *integer ratio* between some fundamental and its harmonics and partially harmonic ones by a *non-integer ratio*, that is $\frac{partial}{f_1}$ is not an integer. The following example is different from the previous one only in line 10, where we introduce randomness to the frequencies of our components. Among other thing, inharmonicity is an important component of many acoustic instruments and, accordingly, it is desired if one wishes to have real instrumental timbres somehow echoed in their synthesizers. Note that new random values are generated each time the code is evaluated.

```
1  // variable envelopes with quasi-integer ratio
2  {
3  Mix.new( Array.fill(50,
4      { arg k ;
5      var incr = 1 ; // quasi-integer. Try increasing to 2, 5, ...
6      var env ;
7      i = k+1 ;
8      env = LFNoise1.ar(LFNoise0.ar(10, add:1.75, mul:0.75), add:0.5, mul:0.5) ;
9      SinOsc.ar(50*i
10         +(i*incr).rand,
11         mul: 0.02/i.asFloat.rand)*env })
12 )}.scope
```

A partially harmonic spectrum can be also obtained employing simple sinusoidal components—80 in the following example:

```
1  // A generic partial spectrum
2  {
3  var num = 80 ;
4  Mix.new( Array.fill(num, { SinOsc.ar(20 + 10000.0.rand, 2pi.rand, 1/num) }) );
5  }.scope
```

It is, then, possible to mix sinusoids at a reduced interval around 500 Hz:

```
1  // Modulation around 500 Hz
2  {
3  Mix.new( Array.fill(20, {
4        SinOsc.ar(500 +
5        LFNoise1.ar(
6              LFNoise1.ar(1, add:1.5, mul:1.5),
7              add:500, mul: 500.0.rand), 0, 0.05) }) );
8  }.scope ;
```

Finally, in the following example we add together signals obtained with a diverse range of techniques (employing SinOsc, Blip, HPF, Dust, Formant) which are all defined in terms of the same fundamental frequency f and a pseudo-randomly controlled amplitude (contained in arr). The signal is then panned into stereo by employing Pan2. Note also that the selection of pitches is also controlled pseudo-randomly. The result is an algorithmic micro-composition.

```
1  // Additive in wide sense:
2  // 4 UGens tuned around a freq
3  // mix and pseudo-random panning
4  {
5      var arr = Array.fill(4, {LFNoise1.ar(1, add:0.15, mul:0.15)}) ;

7      f = LFNoise0.ar(LFNoise0.ar(
8          SinOsc.kr(0.25, 0, 0.75, 1).unipolar.round(0.0625),
9          add:0.95, mul: 0.95),
10         add: 48, mul:12).round.midicps; // 24 semitones, 36-60 MIDI
11     Pan2.ar(
12         Mix.new([
13             SinOsc.ar(f, mul:arr[0]),
14             Blip.ar(f, mul:arr[1]),
15             RLPF.ar(Dust.ar(f*0.2), f, mul:arr[2]),
16             Formant.ar(f,mul:arr[3]),
17         ])
18             , LFNoise1.ar(0.2, mul:1)
19  } }.scope;
```

Additive synthesis is particularly suitable for static sounds, where it is sufficient to define what partials should be present. Building waveforms with edges
(such as in the cases of square, triangular, or sawtooth waves) theoretically requires an infinite number of partials, however. Accordingly SC provides us
with dedicated generators: for example Pulse, to generates quadratic pulses
(including square ones) and Saw, to generated sawtooth like ones[10]. It is then
possible to produce largely aperiodic, albeit not "noisy", signals using a series
of sinusoids that are not harmonically related with each other; the result has a
sort of "liquid"-like quality.

---

[10] There is also Blip.ar(freq, numharm, mul, add) which generates a numharm
number of harmonics with respect to the fundamental frequency freq, all having the same amplitude.

```
1  // Filling the spectrum with sinusoids: fractions of tone
2  (
3  x = {|base = 30|  // 30 MIDI = low rumble
4      var comp = 350; // number of components
5      var res = 0.1; // semitone resolution
6      Mix.new( Array.fill(comp,
7          { arg i;
8              SinOsc.ar(freq: (base+(i*res)).midicps,
9                  phase: 2pi.rand, mul: 4.0/comp )}) // random phase
10 )}.scope ;
11 )
12 // we change the base pitch
13 x.set(\base , 50)
```

In this example, 350 sinusoids are created in parallel (quite of a number...). Their number is determined with comp. Our filling strategy calculates frequencies with respect to midi notation, so that components are distributed with respect to the pitch perceived and not according to physical frequencies. The variable res controls the fraction of a semitone that increments counter i from the base fundamental.

It should be noted that, historically, the main problem in additive synthesis has been to perform all calculations needed. Consider, for example, the amount of computational power needed to compute as signal that includes 20 harmonic components, each with its own envelope. Given the advances in digital technology and the efficiency of SC, demanding additive synthesis is not impossible these days. Yet, it remains a rather expensive way to synthesize complex spectra. For instance, the synthesis of percussive sounds, which are largely characterized by noisy components (i.e. ones that are not harmonically related with each other), would require an incredible number of components and envelopes. Such cases are better addressed by subtractive synthesis.

8.3.2 Subtractive synthesis

In the case of subtractive synthesis, the input signal is usually a complex signal with a substantial spectrum. This signal is then filtered so that only the desired spectral components are emphasized and the unwanted ones are attenuated or removed completely.

In general, a filter is an operation that alters an input signal. Filters typically attenuate or emphasize particular frequency components. Its fundamental arguments design includes: filter type, cut-off frequency, order. The most common types of filter fall in either of those categories: *low-pass, high-pass, band-pass* and *band-reject* (or *notch*), depending on the frequency bands they affect. Figure **8.12** visualizes these basic approaches.

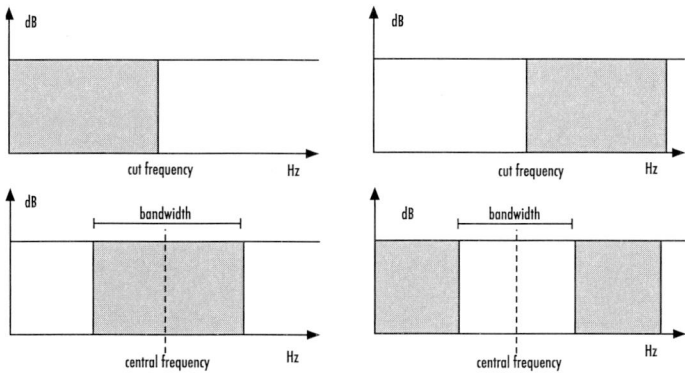

Figure 8.12   Types of filters: low-pass, high-pass, band-pass, band-reject. The grayed area represents the frequency intervals the filters leaves unaffected.

A low-pass or high-pass filter should, ideally, maintain all the frequencies below or above, respectively, a given *cut-off frequency*. In the same way, band-pass or band-reject filters should, ideally, eliminate frequencies that lie outside or inside, respectively, the given frequency band. The latter is typically defined in terms of a *bandwidth* and a *central frequency*: given a spectral frequency, the bandwidth parameter defines an area around the cut-off frequency. For example, a filter that passes all the frequencies in a range of $[100, 110]$ Hz would have a bandwith of 10 Hz and a cut-off frequency of 105 Hz. Note that the filters represented in the figure are *ideal* filters. The difference between an ideal world and the reality is demonstrated with the following example:

```
1 { LPF.ar(WhiteNoise.ar, freq: 1000) }.freqscope ;
```

The LPF UGen is, as the name may suggest, a low pass filter; freq indicates the cut-off frequency. Since the source is white noise (note the patching), when visualized the result should look similar that in Figure **8.12**. In reality, the attenuation is always progressive (or sloped) rather than an abrupt cut-off; the steepness of the attenuation curve is defined by the filter's order. Since there is no known implementation to create ideal filters, the cut-off frequency is defined as the point in the curve where a 3 dB attenuation can be achieved. If the passage from the unaltered to the altered region of the filter is gradual, there is one more argument to consider: the slope of the curve, measured in dB per octave. It is the latter that determines the order of the filter. For example, a first-order filter is characterized by a progressive attenuation of 6dB per octave, a second-order filter should give us a slope of 12vdB/octave, a third-order by 18 dB/octave, etc. LPF UGen implements a second-order low pass filter. We can implement filters of greater orders through a nesting technique; consider the following example:

```
1 { LPF.ar(
2     in: LPF.ar(in:WhiteNoise.ar, freq: 1000),
3     freq: 1000)
4 }.freqscope ;
```

The resulting spectra (from both examples) are illustrated in Figure **8.13**.

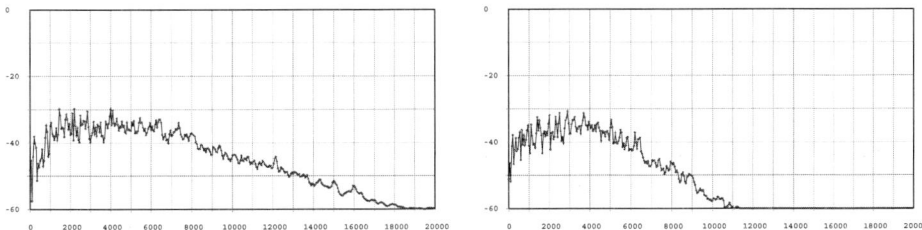

Figure 8.13   LPF filter, spectra: single filter and two cascaded ones.

In the case of white noise we see that all the regions above the cut-off frequency have (ideally) 0 energy. This attenuation reduces the overall volume of the signal in the time domain. Typically, the more consistent the filter, the more the attenuation. Accordingly, when so much energy is lost in the result, we typically raise the amplitude of the output—a process often referred to as *balancing*).

A parameter relevant to band pass filters is that of Q. Q intuitively represents the degree of the filter's resonance. More formally:

$$Q = \frac{f_{cutoff}}{bandwidth}$$

Therefore, Q is the relationship between the cut-off frequency and the bandwidth. Such a factor lets us compensate for the known problem of frequency perception (which, as already explained, is perceived logarithmically rather than linearly). When Q is constant, the bandwidth adjusts with respect to the cut-off frequency so that the perceived size of filter's active band is always identical. For example, if:

$$f_{central} = 105$$
$$f_{high} = 110$$
$$f_{low} = 100$$

then

$$Q = \frac{105}{110-100} = 10,5$$

If Q is kept constant and the cut-off frequency is incremented to $10,500$ Hz, the band's limits become $11,000$ and $10,000$. This way the bandwidth has increased from 10 to 1000 Hz, conforming to the perceived frequency sensation. Therefore, $Q \propto resonance$, because if Q is high enough, the bandwidth is perceptually stretched and resonates at a given frequency. Accordingly, Q indicates the selectness of the filter, that is, its resonance.

Subtractive synthesis is rather efficient, computationally speaking. At the same time it provides a way to model complex acoustic systems. E.g. the body of a guitar can be thought of as a filter that selectively attenuates certain frequencies and emphasizes others. There are several cases of other instrumental models based on the idea of filtering some initial excitation. A flute, for instance, can be thought of as a filter (the pipe) operating on the sound of the breath flow (which can be approximated as white noise). The human voice is also a typical case: in this case, a glottal excitation is filtered by all the parts of the phonetic apparatus, otherwise thought of as the cavities which transform the excitation accordingly. This explains why subtractive synthesis is central to all standard voice simulating methods, from VOSIM to *Linear Predictive Coding* (LPC)[11].

---

[11] These techniques could be also put under the category of physical modeling, see later

In the rather usual case where we deal with multiple filters, these can be placed in either parallel or in series. In the first case all filters simultaneously operate on the same signal (much like what happens with a bank of oscillators). In the second, the output of each filters constitutes the input of the next one. In not strictly accurate terms, subtractive can be thought of as symmetrical to additive synthesis: if additive synthesis is meant to generate complex signals out of simple ones, subtractive is meant to sculpt a spectra out of denser ones. Even if we could use any kind of complex signal as the source of subtractive synthesis, we typically use spectrally dense signals, such as noises that are characterized by rich spectra (sine waves, for instance, would not make much sense, having just one component). Figure **8.14** illustrates the bandwidths of four filters (with central frequencies $400, 800, 1200, 1600$, respectively) that in parallel process a fragment of white noise which is decreased from 1000, to 10, to 1 Hz, transforming the noise into a harmonic sound.

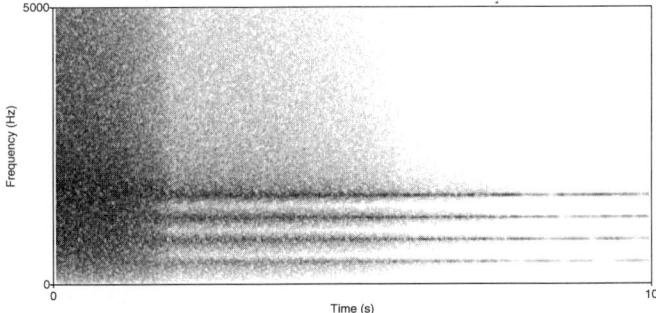

Figure 8.14   Filtering: from white noise to harmonic spectrum.

For reasons related with computational efficiency, SC's standard filters take as an argument the reciprocal of Q, that is rq, rather that Q itself. Accordingly, the smaller the rq argument in the UGen, the narrower the filter. In the following example we implement harmonic filtering in SC: a is an array made up of 10 values that will be used as cut-off frequencies. a is then passed as an argument to a BPF (band pass filter) UGen, which resulting in a 10-channel signal (because of multichannel expansion). In the case of a stereo soundcard, only the first two channels are audible, those that correspond to filtering at 100 and 200 Hz. Employing scope, however, we can visualize all channels and see that the white noise causes each filter to resonate around its cut-off frequency. The following example is similar, but this time the signals are routed into a Mix UGen, resulting in a signal that is subsequently plotted in both the time (with scope) and the frequency (freqscope) domains.

```
1  // Filtering resulting in a harmonic spectrum
2  (
3  var sound = {
4      var rq, i, f, w;
5      i = 10; rq = 0.01; f = 100;      // rq = reciprocal of Q -> bw/cutoff
6      w = WhiteNoise.ar;               // source
7      a = Array.series(i, f, f);
8      // a = [ 100, 200, 300, 400, 500, 600, 700, 800, 900, 1000 ]
9      m = BPF.ar(w, a, rq, i*0.5);
10 } ;
11 sound.scope(10) ;      // see 10 channels of audio, listen to first
12 )

14 (
15 var sound = {
16     var rq, i, f, w;
17     i = 10; rq = 0.01;f = 100;       // rq = reciprocal of Q -> bw/cutoff
18     w = WhiteNoise.ar;               // source
19     a = Array.series(i, f, f);
20     // a = [ 100, 200, 300, 400, 500, 600, 700, 800, 900, 1000 ]
21     n = BPF.ar(w, a, rq, i*0.5);
22     m = Mix.ar(n);                   // mixDown
23 } ;
24 sound.scope ;                 // mixdown: waveform
25 sound.freqscope ;             // mixdown: spectrum
26 )
```

In the next example you can compare several sources and verify the output of the filters. The code also makes use of a bus: the filter synth reads from the bus ~bus where the rest of synths write (remember that these are placed before the former in terms of order of execution).

```
1   // Sources

3   ~bus = Bus.audio(s, 1) ; // bus where to route

5   (
6   // a filtering synth
7   ~filt = {|in|
8       var input = In.ar(in) ; // reads from the in bus
9       var i = 10; q = 0.01; f = 100;
10      a = Array.series(i, f, f);
11      n = BPF.ar(input, a, q, i);
12      m = Mix.ar(n)*0.2;
13      [m, Silent.ar(1), input] // writes on 3 buses to visualize
14  }.scope ;
15  )

17  ~filt.set(\in , ~bus) ; // reads from ~bus

19  // various source synths writing on ~bus
20  ~source = { Out.ar(~bus, Pulse.ar(100, 0.1, mul: 0.1)) }.play
21  ~source.free ; // and are deallocated
22  ~source = { Out.ar(~bus, Dust2.ar(100, mul: 1) ) }.play
23  ~source.free ;
24  ~source = { Out.ar(~bus,LFNoise0.ar(100, 0.1, mul: 1) ) }.play
25  ~source.free ;
26  ~source = { Out.ar(~bus, WhiteNoise.ar(mul: 0.1)  ) }.play
27  ~source.free ;
28  ~source = { Out.ar(~bus, BrownNoise.ar(mul: 0.1)  ) }.play
29  ~source.free ; ~filt.free ;
```

Note that the 3 channels are plotted herein, the first two being the public audio buses—the ones we hear—and the 3rd being the original source—which is shown here for comparison only and is not actually heard on a two channel system. Bus 1 (the second) carries the output of a Silent UGen, which is just a signal with 0 amplitude.

8.3.3 Analysis and resynthesis: Phase vocoder

As already discussed, with both additive and subtractive synthesis, it is of fundamental importance to control the parameters that define the (in)harmonic components of the resulting spectrum. In this respect, additive and subtractive synthesis are analogous, but opposite in their approach. Consider e.g. that several systems can be implemented both in terms of oscillator or filter banks. In the case of the various analysis and resynthesis techniques, the values of the various parameters are set with respect to data derived from the analysis of a pre-existent signal. The process typically comprises three phases:

1. creation of some data structure to hold the analysis data;
2. modification of the analysis data;
3. resynthesis based on the modified data.

There are several ways to implement such an architecture. The case of the Phase Vocoder is particularly interesting and easily implemented in SC. In PV, the signal is typically analyzed with respect to STFT (Short Time Fourier Transform). In STFT the signal is separated in frames (windows, or segments of the signal), each of which is routed to a bank of parallel filters linearly distributed between 0 and sr (sampling rate) frequencies. The result of the analysis for every filter and for every frame determines a series of sinusoidal components—the frequencies of the latter equal the cut-off frequencies of the filter. Essentially:

1. each frame of the original signal is broken down into a set of components that determine values for amplitude and phase.
2. frequency envelopes are constructed with respect to the above values; the amplitude, phase and instantaneous frequency of each sinusoidal component are calculated by interpolating between the values of successive frames.

Accordingly, the envelopes exceed the limits of the single frame and may be used to control a bank of oscillators so that the original signal is reproduced by means of additive synthesis. A particularly efficient implementation of the STFT is the FFT (Fast Fourier Transform).

If the analysis file is not modified, the FFT-based synthesis theoretically reproduces a signal identical to the original. In reality, however, data is always altered/lost in some way. The PV leverages the most interesting aspect of the FFT analysis: the relationship between time and frequency. It is thus possible to change one of the parameters without altering the other. More, it is possible to independently control each individual component or to extract the envelopes of only some of them.

Since analysis is necessary prior to the actual synthesis, and given its being computationally expensive, the classical implementations keeps these two stages distinct: the analysis' results are written to files which can be loaded into memory and re-synthesized at some later time. Real-time implementations are also possible; they would require the allocation of a buffer where the analysis data would be written once calculated form the source. The size of the buffer (that must be a power of 2 for reasons of efficiency) would, then, correspond to the size of the analysis window (the frame). Essentially, every frame drawn from the signal is to be stored in the buffer, replacing the previous one. The data stored in the buffer is the instantaneous spectrum of the frame, which is dealt with as a single time unit. This operation is carried out by the FFT UGen which implements a Fast Fourier Transform on the window. The data, or instantaneous spectrum, stored in its buffer can be then processed according to set of extremely powerful Phase Vocoder UGens (all prefixed with PV_). Phase vocoder operations are typically performed on the same buffer so that they replace the existent spectral data. When all processing is done, the buffer still contains frequency-domain data, and as such it has to be converted back to a time-domain waveform before we can send it to the audio output. This conversion is carried out by IFFT UGen which performs the Inverse Fast Fourier Transform on the signal. The entire schema can be conceptualized this way:

input signal → FFT → PV_... → IFFT → output signal

PV_ elements are optional in theory, in the sense that the analysis data can at any time be re-synthesized directly. Such a scenario is not particularly meaningful; nevertheless, it enables an explanation to the process:

```
1  b = Buffer.read(s, Platform.resourceDir +/+ "sounds/a11wlk01.wav") ;

3  (
4  SynthDef("noOperation", { arg soundBuf, out = 0 ;
5      var in, chain ;
6      var fftBuf = LocalBuf(2048, 1) ;
7      in = PlayBuf.ar(1, soundBuf, loop:1) ;
8      chain = FFT(fftBuf, in) ;       // time --> freq
9      Out.ar(out,
10         IFFT(chain) // freq --> time
11     );
12 }).play(s, [\soundBuf , b]) ;
13 )
```

In this example, the audio file is loaded to the buffer b and read by the synth; the latter converts the former to the frequency domain, via FFT and later re-converts it back to a time-domain waveform, via IFFT. The signal in, which results from the playing back the buffer (via PlayBuf), is converted to a frequency-domain signal via FFT (note the variable assigned to the spectrum is typically named chain to indicate that it refers to a series of successive frames). The analysis data is stored in the fftBuf buffer, created by the convenience LocalBuf UGen. This construct is useful whenever a buffer is contained within a synth, like here[12]. Also note that the buffer's size is 2048, which is the 11th power of 2. Therefore, Out's output is the resultant signal from the inverse conversion of the data from the frequency to the time domain. In an ideal world, there would be no audible loss or artifacts in such a case, and the resultant signal would be identical to the input. In the real world, however, all sorts of artifacts occur which are congenital to the FFT/IFFT operations; such artifacts are typically unnoticeable or bearable, but in certain cases they might result in very noisy signals. Using PV_, the above schema is the basis to a very powerful synthesis technique as demonstrated in the following examples. Consider the following example: the SynthDef implements a "noise gate"; that is, it only lets out input signals that are louder than a given amplitude—herein set interactively with the mouse.

```
1  b = Buffer.read(s, Platform.resourceDir +/+ "sounds/a11wlk01.wav") ;

3  SynthDef(\noiseGate , { arg soundBuf ;
4      var sig;
5      sig = PlayBuf.ar(1, soundBuf, loop:1);
6      sig = sig.abs.thresh(MouseX.kr(0,1)) * sig.sign;
7      Out.ar(0, sig);
8  }).play(s, [\soundBuf , b]) ;

10 // what's going on?
11 s.scope(1) ;
```

----

[12] Note that neither FFT nor IFFT and LocalBuf are invoked with *ar/*kr methods; the latter are only meaningful in the time-domain.

The important part of this code sample is in line 6. The negative part of the signal is converted to positive via abs, therefore thresh lets unaltered all those samples with amplitude greater that the one selected (via MouseX), even if the latter is expressed as a negative number, and substitutes all the rest with 0. In this way all parts of the signal that are inferior to a certain amplitude are silenced. Typically, such an operation is used to eliminated background noises that are represented as a constant low-amplitude bias in the input signal. The output signal, however, would be now a unipolar one, that would sound as if transposed an octave up (because of the folding). The method sign converts all negative values to -1 and all positive to 1. Therefore, multiplying each value of the noise gated signal with the original to which we have applied sign will result in a bipolar noise gated signal. It is worth noting that, as usual with SC, we could get exactly the same result with a different implementation[13]. For example:

```
1   SynthDef("noiseGate2", { arg soundBuf = 0;
2       var pb, ir, mx;
3       mx = MouseX.kr;
4       pb = PlayBuf.ar(1, soundBuf, loop: 1);
5       ir = InRange.ar(pb.abs, mx, 1);
6       Out.ar(0,  pb * ir)
7   }).play(s, [\soundBufnum , b]);

9   // what's going on?
10  s.scope(1) ;
```

The UGen InRange.kr(in, lo, hi) returns 1 for each sample that lies with the $[lo, hi]$, 0 interval and 0 otherwise. In the example, lo is set via MouseX while hi is always 1. Accordingly, the output signal ir is always a sequence of 0 or 1 with respect to the absolute value of the signal: if the absolute value of the signal (abs) is greater than low—here assigned as the variable mx— the output would be 1. The signal will be essentially the same or less than 1, which is the absolute maximum for normalized signals. The original signal pb is then multiplied with ir, which is reset if the amplitude is less that the selected one

---

[13] In SC mailing list, Nathaniel Virgo proposes the first implementation while Stephan Wittwer the following one.

(mx) (being multiplied with 0) and left unaltered if it is greater that the latter (being multiplied with 1).

You will notice that, albeit efficient, the noise gate produces significant "holes" in the signal. The following example, essentially a spectral noise gate, follows a different approach. When applied on a spectrum (chain in this case), the PV_MagAbove will allow the amplitude coefficients of the spectrum (typically referred to as "bins") to pass through unaffected if their value is greater than a threshold (controlled by MouseX), and will silence them if their value is less that the latter. It is possible to completely eliminate components and background noises with respect to those parts of the spectrum that are foregrounded and more spectrally significant. This is a very powerful, but not painless, operation, since certain frequency components might be completely removed this way while still affecting the foreground sounds[14].

```
1  // Spectral gate
2  b = Buffer.read(s, Platform.resourceDir +/+ "sounds/allwlk01.wav") ;

4  (
5  // Mouse control over spectral threshold
6  SynthDef("magAbove", { arg bufnum, soundBuf ;
7      var in, chain;
8      var fftBuf = LocalBuf(2048, 1) ;
9      in = PlayBuf.ar(1, soundBuf, loop: 1);
10     chain = FFT(fftBuf, in);
11     chain = PV_MagAbove(chain, MouseX.kr(0, 40, 0));
12     Out.ar(0, 0.5 * IFFT(chain));
13  }).play(s,[\soundBuf , b]);
14  )
```

Filtering also has a frequency-domain counterpart. Indeed, it is possible to eliminate particular frequency bins in an spectrum—e.g. those above or below a certain frequency threshold, to create low-pass and high-pass filters. In the

---

[14] Note that the threshold is defined by MouseX in terms of "magnitude". The exact definition and the measurement unit of the latter is rather vague in SC's documentation; accordingly we typically set this value empirically since SC lacks tools to tell us what these values should be. It should be noted that the maximum value of a given bin would equal the sum of the points that construct the FFT windowing function.

following example, the mouse controls the wipe argument of a PV_BrickWall UGen; the latter is essentially a frequency-domain low/high-pass filter with a very abrupt slope. When wipe is 0 there is no effect; when it is < 0 the UGen works as a low pass filter; when it is > 0 as a high-pass. Possible values are within the [−1, 1] range, which means that the cut-off frequency need to be estimated linearly in terms of desired frequency and Nyquist frequency (since FFT bins are linear, a wipe of 0.5 at a sample rate of 44, 100 will give us a cutoff around 0.5 × 22, 050 (the Nyquist at this sample rate), or 11025 Hz). The filter is named "brickwall" because of the steepness of its slope, which is essentially a vertical line: all components above the cut-off frequencies are zeroed, rather than attenuated with a rolloff that is characteristic of the filters described earlier in this chapter. In our example, before sending the filtered signal to the output we also use a Normalizer UGen. The latter rescales the signal so that its peak value equals level (here = 1) (i.e. it normalizes it). This way we compensate for any energy loss due to the filtering.

```
1  b = Buffer.read(s, Platform.resourceDir +/+ "sounds/a11wlk01.wav") ;
2  s.freqscope ; // what's going on?
3  (
4  // FFT filter
5  SynthDef("brickWall", { arg soundBuf ;
6      var in, chain;
7      var fftBuf = LocalBuf(2048, 1) ;
8      in = PlayBuf.ar(1, soundBuf, loop: 1);
9      chain = FFT(fftBuf, in);
10     chain = PV_BrickWall(chain, MouseX.kr(-1.0,1.0, 0));
11     // -1.0 --> 0.0: LoPass ; 0.0 --> 1.0: HiPass
12     Out.ar(0, Normalizer.ar(IFFT(chain), level:1));
13 }).play(s, [\soundBuf , b]);
14 )
```

PV UGens enable us to process frequency without regards to time. Amongst the available UGens one can find PV_BinShift which is used in the following example. PV_BinShift(buffer, stretch, shift) translates and scales bins with respect to a shift and a stretch, respectively: for example, given a three-component spectrum [100, 340, 450], a translation (shift) by +30 results in a new spectrum [130, 370, 480]. In the following piece of code, the stretch argument is 1 and the shift is controlled by MouseX which produces values within the

$[-128, 128]$ range. Observe how the entire spectrum is shifted left or right according to the way the mouse moves.

```
1  b = Buffer.read(s, Platform.resourceDir +/+ "sounds/a11w1k01.wav") ;
2  // what's going on?
3  s.freqscope ;
4  s.scope ;

6  (
7  SynthDef(\fftShift , {  arg soundBuf ;
8      var in, chain;
9      var fftBuf = LocalBuf(2048, 1) ;
10     in = PlayBuf.ar(1, soundBuf, loop: 1);
11     chain = FFT(fftBuf, in);
12     chain = PV_BinShift(chain, 1, MouseX.kr(-128, 128) );
13     Out.ar(0, 0.5 * IFFT(chain).dup);
14 }).play(s, [\soundBuf , b]);
15 )
```

In a similar manner, we can scale the values of bins via the stretch argument: as illustrated with the following example (in particular when moving the mouse from left to right), variations of scale within the interval $[0.25, 4]$ cause the entire spectrum to expand or contract while keeping the same ratio between the component's of the spectrum.

```
1  SynthDef("fftStretch", {  arg soundBuf ;
2      var in, chain;
3      var fftBuf = LocalBuf(2048, 1) ;
4      in = PlayBuf.ar(1, soundBuf, loop: 1);
5      chain = FFT(fftBuf, in);
6      chain = PV_BinShift(chain, MouseX.kr(0.25, 4, \exponential ) );
7      Out.ar(0, 0.5 * IFFT(chain).dup);
8  }).play(s, [\soundBuf , b]);
```

Two final notes.

The dup(n) method, defined on Object which is the superclass of all classes in SC, returns an array that comprises n copies of that object. The default n value is 2. In the previous example, dup has been invoked on IFFT UGen, which returns an [IFFT,IFFT] array. In this way, we use the array to create a multi-channel expansion for a stereo signal.

Finally, it has to be noted that the values of the arguments of the various PV_ UGens are typically not expressed in Hz. They are sometimes expressed in some normalized form (in the case e.g. of PV_BrickWall) and some others in implementation-dependent ranges (e.g. in the case of PV_BinShift). As already discussed, in the latter cases it is not always clear what the end result will be, and users are encouraged to play with these parameters and observe what happens based on intuition or empirical estimations. But a word of caution: these results will be sample-rate dependent!

8.4 Physical Modeling

Physical modeling techniques are intended as simulations of sound-generating physical processes. The physical modeling paradigm aims at generating audio by means of a series of equations which describe the physical behavior (in acoustic and / or mechanical terms) of real world sound-producing phenomena. Where with additive synthesis we would approximate the timbre of a clarinet by means of a suitable conglomeration of sine waves to represent the odd harmonics of some fundamental frequency, with physical modeling we make a mathematical model that describes the operation of a real clarinet's physical components. The latter would, then, be described in respect to a series of arguments such as the size of the tube, the behavior of the reed, the density of the wood, etc.

Physical modeling techniques are computationally expensive, typically, because of the number of complex calculations that are necessary to resolve linear equations. In SC, physical modeling is typically implemented within dedicated UGens using primitive objects. The variety of such UGens that are available with the official distribution is rather limited, but there several others available as external extensions. The theoretical background of physical modeling is beyond the scope of this book—interested users should consult other resources that are dedicated to the subject. The following two examples are, nevertheless, useful as an informal introduction to physical modeling synthesis in SC.

A physical model is a formal representation of an acoustic system. As a rule of thumb, there are several ways to formalize acoustic systems. For example, many acoustic devices can be described as systems of filters arranged in various combinations. The degree of precision in which the behavior of the filters is described determines the accuracy of the modeling. From this point of view, subtractive synthesis is a physical modeling technique.The human vocal apparatus, for instance, produces the vowels by means of modulating the structure of a pipe complex that is composed of multiple components (oto-pharyngeal conduct, oral and nasal cavity, etc), in order to control the way an acoustic source is filtered. Phonetics shows that a key feature in the organization of vowels is the presence of formants, i.e. of spectral bands in which the energy is concentrated. The spectrum of each vowel includes from 4 to 5 formants: typically, the first two are used to linguistically recognize the vowel, and the others to identify the speaker. Note that the formants are independent of the fundamental frequency: in this way, an "a" pronounced by a child is recognized precisely as such even if the fundamental frequency is much more acute than that of the phonation of an adult. Phonetic studies define a two-dimensional space in which the axes represent the first two formants. On top of it, points (in reality, areas) representing the various vowels that are defined for various languages. Each language features a different number of important points which typically occupy different areas in the diagram. Figure **8.15** illustrates the vowel space for Italian.

Digital voice synthesis is a difficult research topic. As an exercise, we introduce a simplified "cartoonified" model for vowels based on the two-formant space. In the following example the source is a sawtooth wave, which, up to a certain extent, resemble the acoustic signal produced by the glottis. The frequency is determined by fund, the value of which is predefined as 70 Hz, a typical value for a male speaker. Then, the source signal is filtered by two band-pass filters (in parallel) with respect to those two frequencies that represent the formants: these are, f1 and f2. To compensate for the loss of energy, the outputs of the filters are normalized before being mixed together.

---

[15] From Ferrero, Genre, Böe, Contini, *Nozioni di fonetica acustica*, Torino, Omega, 1978.

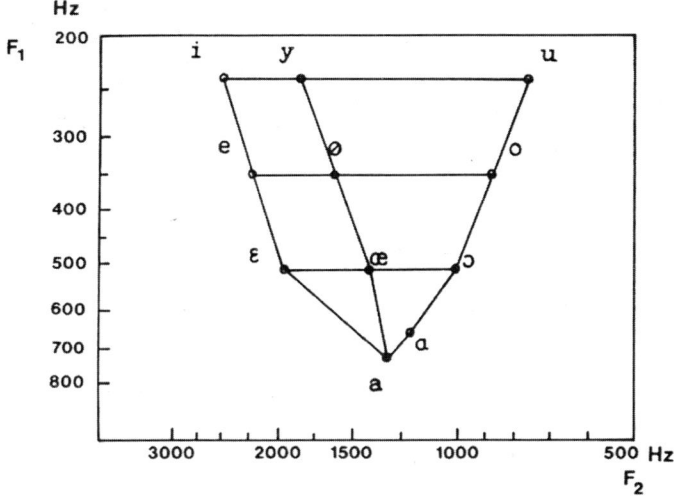

Figure 8.15 Vowel space for Italian[15].

```
1  // A generator for vowel spectra
2  SynthDef(\vocali , { arg f1, f2, fund = 70, amp = 0.25 ;
3      var source = Saw.ar(fund); // source
4      var vowel =
5        Normalizer.ar(BPF.ar(source, f1, 0.1))
6        +
7        Normalizer.ar(BPF.ar(source, f2, 0.1))
8        * amp ; // general vol
9      Out.ar(0, vowel.dup)
10 }).add ;
```

One of the key points in physical modeling is control. Every physical model prerequisites the definition of a series of arguments with respect to which the former is defined. In absence of such arguments, even the best model is useless. The following example features a graphic interface that defines a two-dimensional control space where formants can be expressed in Hz.

```
1  (
2  ~synth = Synth(\vocali ) ;

4  d = 600; e = 400;
5  w = Window("Formant space", Rect(100, 100, d+20, e+20) ).front ;
6  Array.series(21, 2500, 100.neg).do{|i,j|
7      StaticText(w, Rect(j*(d/21)+5, 10, 30,10 ))
8      .font_(Font("Helvetica", 8))
9      .string_(i.asString)
10 } ;
11 Array.series(14, 200, 50).do{|i,j|
12     StaticText(w, Rect(d, j*(e/14)+20, 30, 10 ))
13     .font_(Font("Helvetica", 8))
14     .string_(i.asString)
15 } ;
16 u = UserView(w, Rect(0, 20, d, e)).background_(Color.white) ;

18 ~vow = (
19     // data for "sensitive" points in Hz
20     \i :[2300, 300], \e : [2150, 440], \E : [1830, 580],
21     \a : [1620, 780], \O : [900, 580], \o : [730, 440],
22     \u : [780, 290],\y : [1750, 300],\oe : [1600, 440],
23     \OE : [1400, 580]
24 ) ;
25 f = {|v, f2, f1| StaticText(u,
26     Rect(f2.linlin(500, 2500, d, 0),
27         f1.linlin(200, 800, 0, e)-18, 40,40))
28     .string_(v).font_(Font("Helvetica", 18))
29 } ;
30 ~vow.keys.asArray.do{|key|
31     var f2 = ~vow[key][0] ;
32     var f1 = ~vow[key][1] ;
33     f.value(key.asString, f2, f1)
34 } ;

36 w.acceptsMouseOver = true ;
37 u.mouseOverAction_({|v,x,y|
38     ~synth.set (
39         \f2 , x.linlin(0, d, 2500, 500).postln,
40         \f1 , y.linlin(0, e, 200, 850).postln,
41     )
42     })
43 )
```

The code is purposely a bit "dirty", to illustrate how real-life prototypes often look. For example, a number of environment variables are used for brevity (d,e,f,w,u) which should have more meaningful declared variable names. In line 2 the synth ~synth is instantiated. The variable d and e determine the length and width of the window w which contains the UserView (a user-controlled graphic display). The two arrays (lines 6–15) only serve to generate labels of the axes. Lines 36-42 define the actions related to user interaction via mouse. First we need to indicate that all the children of the window w accept mouse events (line 36), then we need to specify what happens when the mouse is moved on the view u. As can be seen, the method mouseOverAction enables access the horizontal (x) and vertical (y) coordinates of the mouse's positioning (in pixels) within the view. They are used to set the value of $f1$ and $f2$, respectively, once scaled. Function f is the meant to label the space. This makes it possible to generate a set of labels (lines 30-34) from the data ~vow.

Another technique that is typically associated with physical modeling is the so-called Karplus-Strong algorithm. This technique is similar to the "waveguide" technique, which simulates the passage of a source signal in a physical device in terms of a sequence of filters and delays that are supposed to simulate the reflections of the waves across various acoustic surfaces. Both techniques revolve around filters and delays, however, while the waveguide paradigm is supposed to be a resonator's physical model, the Karplus-Strong is a much more abstract algorithm. Even a simple implementation, however, allow us to mimic the behavior of strings or pitched percussion instruments. As can be seen in Figure **8.16**, a noise generator fills a table (such as those encountered in the case of the wavetable synthesis) once for each event or note: at this point the generator is disconnected and data is received from the modifier. The latter reads the first value from the table, sends it to its output and also adds it to the top of the table through a delay line. This way the table is being constantly rewritten for the duration of the event: this is known as a "recirculating wavetable". For example, with a table of 256 points and a sampling rate of 44,100, the table is read (and edited) about 172 times a second ($\frac{sr}{n_{tab}} = \frac{44.100}{256} = 172.265$).

Changes can be very rapid because the table is being rewritten many times per second. The gist of this operation is the particular way each value is modified before it is rewritten into the table: with appropriate filtering various timbres can be obtained; plucked strings, for instance, have an intense attack which decays rapidly both in amplitude and spectral richness. There are two consequences in rewriting the processed sample back to the table. First, the resulting signal is not necessarily a noise; unlike the pseudo-random sequence of values

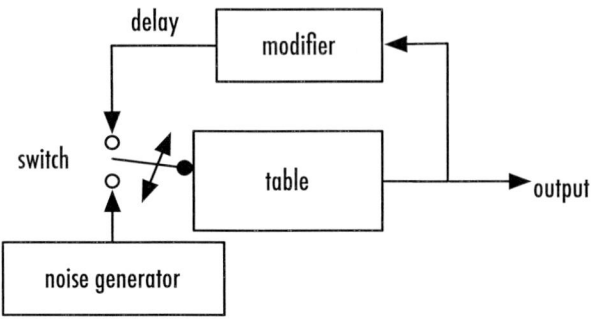

Figure 8.16   Karplus-Strong algorithm.

that were initially placed into the table may suggest, it can be tuned sound as well. Indeed, the resulting signal is rather periodic because the table is read repeatedly over the given period and because its values will remain constant after a number of cycles (this is similar to the design of digital oscillators; in the above example, the frequency of the resulting waveform would be approximately 172 Hz). Secondly, it is possible to easily simulate the dynamic and spectral envelopes of pitched strings. The attack portion corresponds to the initial emission of pseudo-random numbers from the wavetable, which produces a rather diffused spectrum. SC implements this algorithm in the Pluck UGen.

```
1  { Pluck.ar(
2        in:WhiteNoise.ar(0.25),
3        trig:Impulse.kr(1),
4        delaytime:60.midicps.reciprocal,
5        maxdelaytime:60.midicps.reciprocal,
6        decaytime:MouseY.kr(1,20,1),
7        coef:MouseX.kr(-0.75, 0.75))
8  }.play ;
```

As seen in the example, the parameters include the excitation source that fills the table (here a white noise) and a trigger that indicates precisely the filling time (here given by the rate of an Impulse UGen). The frequency is controlled by specifying the delay time, which becomes the period of the desired frequency (specified in terms of the reciprocalof the 60.midicps frequency). The argument maxdelaytime determines the size of the internal buffer, which must be equal to or greater than the period of the frequency sought. The two most interesting parameters are the decay time which is an indicator of the resonance (expressed as the attenuation of 60 dB in seconds) and coeff, which controls the internal filter applied to the values that are rewritten in the table. You may notice that this affects how our perception of "pitched" the string is can be altered with these controls. An interesting exercise can be the implementation of systems inspired by KS algorithm.

```
1  (
2  // Karplus-Strong
3  SynthDef(\ks , {
4      arg freq = 440, amp = 1, out = 0, thresh = -90, decrease = -0.25 ;
5      var baseFreq = 48.midicps ; // base freq, arbitrary
6      var buf, index, sig, num = 2, scale = decrease.dbamp ;
7      var samples = 44100/baseFreq ;
8      var actualValue ;
9      buf = LocalBuf(samples) ;
10     // random table = white noise
11     buf.set(Array.fill(buf.numFrames, { 2.0.rand-1 }));
12     index = Phasor.ar(
13         trig:Impulse.ar(buf.numFrames/SampleRate.ir),
14         rate:freq/baseFreq,
15         // -> reading rate
16         start:0, end:buf.numFrames, resetPos:0);
17     actualValue = BufRd.ar(1, buf, index) ;
18     // circular reading, until signal goes under threshold
19     Out.ar(out, actualValue*amp) ;
20     DetectSilence.ar(actualValue, thresh.dbamp, 0.1, 2) ;
21     // circular rewriting
22     sig = Array.fill(num, {|i| BufRd.ar(1, buf, index-i)}).sum/num*scale;
23     BufWr.ar(sig, buf, index) ;
24  }).add ;
25  )

27  Synth(\ks , [\freq , 50.midicps, \amp , 0.5, \out ,0 ]) ;
```

The SynthDef is based on an arbitrary reference frequency, defined as base-Freq (a low C) and assumes that the sampling rate is $44,100$. Then, a local buffer buf is allocated with a size equal to $\frac{44,100}{baseFreq}$ samples. In other words, the buffer's size equals the period of the reference frequency. The buffer is, then, filled with pseudo-random values in the $[-1.0, 1.0]$ range which essentially represents an initial white-noise signal. The problem now is how to read the table. The Phasor UGen generates a signal which suits this purpose. Specifically, it is a ramp signal that progressively increments in values that can be used as indexes into the buffer. Here, Phasor is triggered by an Impulse UGen, the period of which corresponds to the buffer's size. In practice, every time the buffer reaches its end, Phasor is re-triggered to read it again from the beginning (from position 0, as determined by the resetPos argument). The rate argument indicates the speed according to which the buffer is read and is calculated in terms of the ratio between the desired freq and the reference frequencies. If freq is two times baseFreq, the buffer will be read two times faster. Note that Phasor operates at audio rate (*ar), so that with every sample a new index is calculated.

The variable index is used (line 18) to read the relevant sample (assigned as actualValue) from the buffer with respect to BufRD UGen, and the resulting signal is sent to the output. The amplitude of the signal is then checked using DetetSilence: if the amplitude is below a certain threshold thresh the synth is deallocated (doneAction:2). So far we have merely read the buffer. The last two lines of the code is where the values are actually modified. A sig value is calculated as the sum of the values contained between index-i and i divided by the total number of samples and scaled by scale. With $num = 2$, just the current and the previous sample are considered. With 0.5 and 1 and scale set to 0.97162795157711 (remember that the predefined dB value is -0.25 and is converted into linear amplitude): then $\frac{0.5+1.0}{2} \times 0.97162795157711 = 0.72872096368283$. The new value is then written in the place of the previous (1) which results in an attenuation effect. The diagram of the SynthDef is illustrated in Figure **8.17** (part of the buffer is not shown).

In this example, the modifying operation has essentially been an "averaging filter" implementation of a low-pass filter: the value of the output sample is obtained as the *average* of the current sample and the one that precedes it. Obviously, such a filter would cause a progressive attenuation of the high-frequency components in the input signal. When circulating such a signal through the buffer, the result is a low-pass filter that progressively eliminates the spectral richness of the initial white noise input. The resulting sound is characterized by a complex spectrum on its attack phase which progressively fades and tunes to a rather pitched one.

8.5 Time-based methods

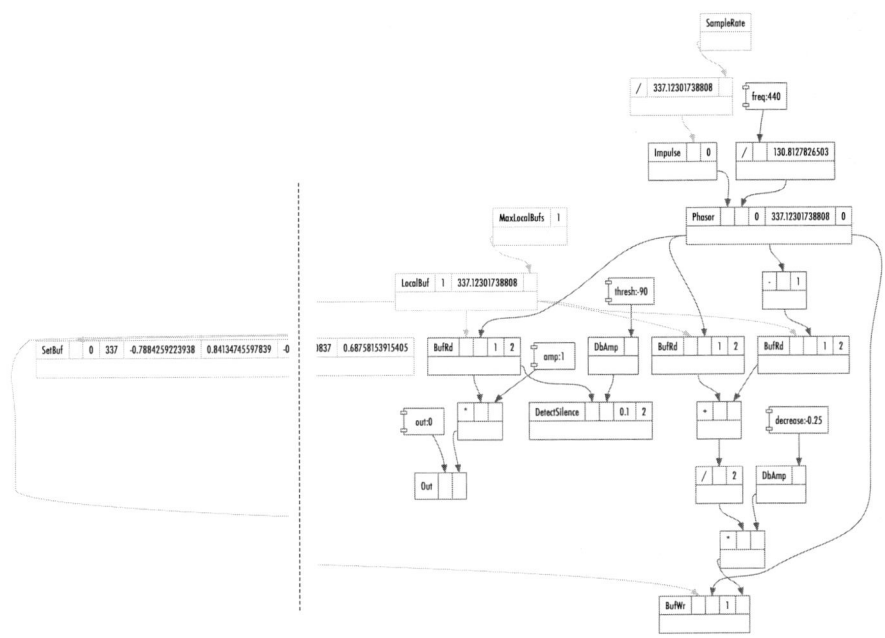

Figure 8.17   Diagram of the ks SynthDef.

In this category we ascribe a set of methods that deal with time domain signal representations. Such methods are typically computational and are often implemented at very high temporal resolutions, ones that cannot be managed by analogue systems. To boot, time-based signals often exploit the very discrete nature of digital audio in various ways.

8.5.1   Granular synthesis

From a granular synthesis point of view, sound is not just a waveform but also a "corpuscle". The usual metaphor is that of pointillism: a series of concatenated microscopic sonic fragments that are perceived as a single continuous sound, in the same way that the adjacent dots of various colors are perceived as a single hue. Grains are sound fragments with a duration between 1 and 100 milliseconds. Each grain is characterized by a particular amplitude envelope, which greatly affects the resulting sound. Given the number of grains typically required to synthesize some sound, their various properties (e.g. duration, frequency, envelope, waveform) need to be defined in some automated manner. Stochastic methods are often used so that the user determines average, rather than absolute, values for the various parameters. Additionally, it is necessary to define some higher level compositional model: not surprisingly, granular synthesis is a technique most often encountered within an algorithmic composition context. For instance, it is possible to organize the mass of a grain-cloud following the distribution laws of ideal gasses (like Iannis Xenakis did), or by employing clusters that simulate meteorological phenomena such as the structure of clouds (like Curtis Roads did), or finally, with respect to "tendency masks" that govern the properties of grain streams (like Barry Truax did). In general, granular synthesis is computationally expensive, given that the properties of hundreds, if not thousands, of individual grains have to be controlled simultaneously. Accordingly, there are numerous ways to implement granular synthesis: even if they may differ significantly in scope, all techniques that revolve around quasi-impulsive signals can be thought of as granular ones.

```
1  (
2  {
3  // granular synchronous synthesis
4  var baseFreq = MouseX.kr(50, 120, 1).midicps ;
5  var disp = MouseY.kr   ;
6  var strata = 30 ;
7  var minDur = 0.05, maxDur = 0.1 ;

9  Mix.fill(strata,
10     {
11     // source
12     SinOsc.ar(
13         freq: baseFreq +
14             LFNoise0.kr(20)
15             .linlin(-1.0,1.0, baseFreq*disp*1.neg, baseFreq*disp),
16         mul: 1/strata)
17     // envelope
18     * LFPulse.kr(
19             freq:
20             LFNoise0.kr(20)
21             .linlin(-1.0,1.0, minDur.reciprocal, maxDur.reciprocal))
22 })
23 }.freqscope
24 )
```

The above example demonstrates a synchronous approach to granular synthesis. It employs a sine-wave with a square-wave amplitude envelope (LF-Pulse); the latter generates a unipolar signal (within a $[0.0, 1.0]$ range): when the amplitude is > 0, the signal passes through, otherwise it it silenced, resulting in a "windowing" of the signal. The size of each window depends on the cycle of LFPulse, which is determined by minDur and maxDur. Since the duration defines the period of the window, LFPulse's frequency will be $\frac{1}{T}$. In this case, the envelope of the grain is given by the positive cycle of the square wave.

As seen, a number of strata signals are mixed using Mix: the frequency of each signal is given by the addition of baseFreq (controlled by MouseX) with a pseudo-random value between 0 and disp (controlled by MouseY) generated by LFNoise0; this frequency represents a normalized percentage of the base frequency (if $baseFreq = 100$ and $disp = 0.5$, then the oscillator's frequency will vary in the $[100 - 50, 100 + 50]$ range). Note that each SinOsc is associated with a unique LFNoise0.

The former approach can be implemented in real time, if some continuous input is "windowed" appropriately by employing some pulse signal. Note that the generation of the pseudo-random signals has to be taken care by LFNoise0 which requires a refresh rate; here, the latter is arbitrarily set to 20 Hz herein.

Another possible approach would be to think of every individual grain as a "composed" event. That is to say that each grain has to be scheduled procedurally. Such an approach is rather expensive computationally due to the high number of grain-events that need to be scheduled.

Granular synthesis techniques are not necessarily used for *ad nihilo* synthesis; they are often used in order to process some input signal—an operation referred to as "granulation". Granulation can be used to decompose an input signal into grains which can then be re-constructed in all sorts of way. Such an operation necessarily introduces some latency: we first need a proper part of the actual input signal before we can granulate it. Accordingly, real time implementations typically revolve around the use of a buffer where part of the input is recorded.

```
1  b = Buffer.read(s, Platform.resourceDir +/+ "sounds/a11wlk01.wav") ;

3  SynthDef(\grainBuf , { arg sndbuf;
4      Out.ar(0,
5          GrainBuf.ar(2,
6              trigger: Impulse.kr(MouseX.kr(10, 20)),
7              dur: MouseY.kr(0.01, 0.1),
8              sndbuf: sndbuf,
9              rate: LFNoise1.kr.range(0.5, 2),
10             pos: LFNoise2.kr(0.1).range(0.0, 1.0),
11             interp:1,
12             pan:LFNoise1.kr(3)
13         ))
14 }).add ;

16 x = Synth(\grainBuf , [\sndbuf , b]) ;
```

The former example is a simplified version of the example found in the GrainBuf's help file. The buffer b keeps part of the input signal and Grain-Buf granulates it accordingly. The syntax of the latter is explicit in the code. The first argument specifies the number of channels to output, and each grain will be panned according to the signal in the pan argument; unlike other buffer playback UGens, this UGen can only process mono signals directly. The second argument is a triggering signal; each trigger produces a new grain: in the example, this is a train of impulses with a frequency between 10 and 20 Hz. The argument dur determines each grain's duration (herein, between 0.01 and 0.1, that is 10 and 100 ms). The subsequent argument specifies the buffer used (herein b). The arguments rate, pos, interp, pan, envbufnum control the buffer's reading rate (like in the case of PlayBuf), the position of reading (normalized between 0, for the start of the buffer, and 1.0, for its end), the interpolation method, the position in the stereo image (such as in the case of Pan2 and as far as stereo signals are concerned) and, optionally, a buffer with an envelope for each grain.

Several of those arguments are controlled with low-frequency (LF) kind of generators. The method range enables us to scale a signal (assumed to be in the $[-1, 1]$ range) to the interval specified by the two arguments passed: this is more intuitive that using mul and add[16]. The LFNoise2 UGen produces a low-frequency noise with quadratic interpolation between the output values, unlike LFNoise1 which has linear interpolation. This results is a more rounded wave. These signals are polled once when a trigger indicates a new grain should be created.

8.5.2 Techniques based on the direct generation of the waveform

It is worth closing with a brief overview of one of the most useful or the most historically important techniques that revolve around the direct computation of the digital values that constitute a signal. There have been relevant examples since the 1960s (by Gottfried Michael Koenig, Iannis Xenakis, Herbert Brün), which mostly take a modeling approach. In those cases, rather that attempting to simulate some acoustic or other model, we directly manipulate streams of raw numbers. Such approaches could be though of as "constructivist". The signal could be thought of as a sequence of straight line segments, each comprised of n samples, that are algorithmically joined. Individual operations can be thought of in terms of inversion, permutation and transformation of sample blocks. If there is a common thread to these approaches, this would be a

---

[16] Since range expects the signal to be in the $[-1, 1]$ range, it cannot be used in conjunction with mul / add.

willingness to explore digital signals in the way that these are represented by computers. The resulting signals are often rich in asperities, in terms of both spectral components and temporal discontinuities. In the chapter devoted to the fundamentals of synthesis, we discussed a relevant example: the "permutative distortion", where blocks of samples are shuffled so that the the sample sequence [a,b,c,d] becomes [b,a,d,c]. The real-time implementation of a similar instance faces an inherent limitation of SC (and of virtually all audio synthesis software environments, really): the server can not access individual samples—i.e. it is not sample-accurate. The server only deals with the output of UGens which output blocks of samples, rather than individual samples[17]. Therefore, the algorithms described in this chapter cannot be implemented server-side. A possible implementation inspired by the patching distortion technique would be the following[18]:

---

[17] Actually, it is to some extent possible to implement sample-accurate synthesis in SC; for example by means of exploiting LocalIn and LocalOut UGens with a block size of 1 for the control period, or using DbufRd and DbufWr UGens which, given certain conditions, also lets us write/read individual samples.

[18] An even more comprehensive and efficient implementation would involve writing a dedicated UGen, see Giacomo Valenti, Andrea Valle, Antonio Servetti, *Permutation synthesis*, Proceedings XX CIM - Colloquio di Informatica Musicale, Rome 2014.

```
1   (
2   b = Buffer.alloc(s, s.sampleRate * 1.0, 1); // buffer mono, 1 sec

4   SynthDef(\sin , {|freq = 100, buf|
5       RecordBuf.ar(SinOsc.ar(freq), buf, loop:1)}).add ;

7   SynthDef(\perm , { arg buf, permFreq = 10 ;
8       var trig = Impulse.ar(permFreq) ;
9       var startPos = LFSaw.ar(permFreq, iphase:1,mul:0.5, add:0.5); // 0 - 1
10      var periodInSamples = permFreq.reciprocal*SampleRate.ir ;
11      var sig =
12      BufRd.ar(1, buf, Phasor.ar(trig, 1,
13          startPos*periodInSamples,
14          startPos*periodInSamples*2, startPos*periodInSamples)
15      ) ;
16      Out.ar(0, sig) ;
17  }).add ;
18  )

20  // writing on a circular buffer
21  x = Synth(\sin , [\buf , b])
22  // after
23  y = Synth.after(x, \perm , [\buf , b]) ;

25  // control
26  y.set(\permFreq , 20) ;
27  x.set(\freq , 1000) ;
28  // what's going on?
29  s.scope ;
```

Initially we allocate a buffer b, which contains a 1-second signal that is synced to the sampling rate of the server. The buffer will be used to record a signal and make it available for further processing. In order to exchange two sets of samples [a, b] in real time, we must register them as two blocks, reverse them then send them to an output—hence the usefulness of the b used by the synth x (line 21). The SynthDef uses RecordBuf, a UGen that takes the following arguments: an input signal (herein generated by SinOsc), a buffer where the signal should be recorded (b) and loop, which indicates whether the buffer should be read in a circular manner (1) or not (0). In principle, our synth continuously records the signal on a circular buffer. Note that there is no Out in our SynthDef. The generated signal is simply recorded on the buffer: it is not routed to an audio bus. The idea is to record the sine-wave so that we may subsequently manipulate it. Recording, of course, introduces latency which depends upon the size of the buffer at the time when the parameters of SinOsc are set. The modified signal is available on b only after recording.

The following SynthDef follows a permutation-driven approach. Control is implemented in terms of frequency, rather than in terms of number of samples. The LFSaw UGen generates a ramp which is normalized to a unipolar signal in the $[0, 1.0]$ range through mul and add. More, the iphase sets the initial phase of the signal to 1: this means that the UGen starts from after the first half of the cycle. Accordingly, it takes into account the amplitudes from 0.5 to 1, then from 0.0 to 0.5. The frequency is that of the desired permutation, permFreq. For convenience, the period is calculated according to the latter, periodInSamples. At this point, the buffer buf (b in this case) is read using Phasor. As each trig received, the latter advances the reading by 1 sample (2nd argument). As seen, these two are calculated with respect to the multiplication of startPos (where the value is a result of the signal from LFSaw) for the desired permutation period. Then it is read from startPos and until the next period. The final argument of Phasor indicates the reset point each time a trig is received. In line 21 a sinusoid is written on the buffer, then we instantiate the synth that reads from this buffer. Finally, we can control the parameters of the sinusoid and the entire permutation process.

The entire process is shown in Figure **8.18**.

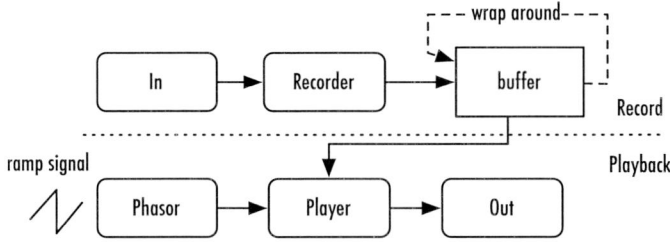

Figure 8.18   Permutation Synthesis.

This way, it is possible to generate complex spectra using simple sinuoids (of course other signals can be also used) and while eschewing any references to acoustics but, rather, simply operating on the temporal structure of the input signal.

8.6 Conclusions

Audio synthesis in the digital domain is an open field to creativity. Accordingly, the possibilities for composers and sound-designers are vast. The examples discussed in this chapter are just a small taste of what audio synthesis in general, and SC in particular, may offer. The most interesting aspect of such techniques is, probably, the way they may be controlled. It is often necessary to incorporate them in broader compositional contexts. This brings us back to what was already discussed as the most important aspect of sound synthesis: the control of audio signals and the control of the overall organization of the sound material.

9 Communication

This chapter aims to quickly discuss the problem of communication in Super-Collider. Specifically, communication as it is defined between some agents. As we said from the beginning, SC immediately identifies two agents, a client and a server. Yet, we have not discussed the details of their communication so far, because it has been conveniently conveniently made transparent by the layer provided by the SC language. However, there are many cases in which to some extent it is necessary to explicitly define a mode of communication between different objects and SC. After several examples of communication between the client and the server, the chapter will provide some preliminary indications with respect to communication between SC and the "outside": for the latter cases, the discussed examples are functional but not entirely autonomous since they often refer to formats, services and applications that are external to SC.

9.1 From server to client: use of control buses

Communication as discussed so far concerned the exchange of messages between client and server. In an expression such as s.boot, the interpreter is the client that converts the SC syntax in a message to the server indicating the latter to start. A similar situation applies to all those constructs that control aspects of the audio synthesis (on server side): the communication is one-way, from the client to the server. Note that the client defers all the calculation of the audio to the server, and the former has no information on the activities of the latter. However, in many cases it is indeed very helpful to have information on what is going on in the audio. A trivial but practical case is plotting inside a GUI window the content of audio signals. If the signal is calculated on the server side, how can it be accessed by the GUI classes that reside on the client side? The solution lies in communication that is the reverse of the model we have seen insofar, where information is sent from the server to the client. The first way in which such a communication can be implemented and utilized in SC is based on the use of control buses. In the following example, the SynthDef is designed as a synthesizer which generates a control signal through a noise generator which interpolates quadratically between successive pseudo-random values, LFNoise2: the result is a particularly smoothed noisy signal. Lines 8-10 create a synth x from the SynthDef, which is routed to the control bus ~bus. The synth y reads from the bus the values generated by x which are then written on the bus. The synth y re-maps these values to the frequency of a sine wave. All this communication, even if linguistically described on the client side, happens between the synths via the bus, that is, on the server side. The next block (12-31) retrieves the values of the control signal from the bus to plot them in a GUI. The heart of the operation is a routine r that every 20 ms (0.05.wait, 29) uses the method get, available for ~bus. This method asks the server to return the client the value that is written on the bus, that is, the signal generated by x. The argument of get represents the value of the bus. Note that get receives a function because it is *asynchronous*: in other words, it asks a value to the server, but it may not know exactly when it will arrive. Therefore, the function is evaluated only when the server has answered the call. In this case, the (shared) environment variable v stores the value from the bus, the counter is incremented, and then the UserView u is updated. Note that the defer method is necessary because the routine r is scheduled (implicitly) on SystemClock. The UserView u has an associated function drawFunc which draws a circle in which the position x depends on a modulo 500, which represents the width of the window u: when this values is exceeded, the counter is reset to zero, and the next circle will be on the 0 point of the x axis. The value of y depends instead on the value of the control signal, v, which is updated through each call to ~bus.get. Through u.clearOnRefresh_(false), the window will not be cleaned up, and then we see the overlapping of the plottings once i is more than 500.

```
1   (
2   SynthDef(\cntr , {arg out = 0, dbGain = 0;
3       Out.kr(out, LFNoise2.kr(0.5))
4   }).add ;
5   )

7   (
8   ~bus = Bus.control(s,1) ;
9   x = Synth(\cntr , [\out , ~bus]) ;
10  y = {SinOsc.ar(In.kr(~bus).linlin(-1, 1, 48, 72).midicps, mul:0.2)}.play ;

12  i = 0;  v = 0.0  ;
13  w = Window.new("Plotter", Rect(10, 10, 500, 500)).front ;
14  u = UserView(w, Rect(0,0, 500, 500)).background_(Color.grey(0.75)) ;
15  u.drawFunc_{
16      Pen.fillColor = Color.red(0.5);
17      Pen.addOval(Rect((i%500), v.linlin(-1.0,1.0, 500,0), 2,2)) ;
18      Pen.fill
19  } ;
20  u.clearOnRefresh_(false) ;

22  r = {
23      inf.do{
24          ~bus.get({ arg amp ;
25              v = amp ;
26              i = i+1 ;
27              {u.refresh}.defer
28          });
29          0.05.wait ;
30      }
31  }.fork
32  )
```

The following example uses a similar mechanism to display the signal amplitude. The UGen Amplitude returns a control signal that estimates the amplitude of the input signal (it is an "amplitude follower"). The control signal can be used directly on the server via patching with other UGens. In the following example, the synth a is an amplitude follower that reads an input audio signal from the bus ~audio and writes the audio signal obtained from the analysis on the control bus ~ampBus. The next block defines a window. Through the function drawFunc, we draw a circle in the center, its size and color depending on four environmental variables (~dim, ~hue, ~sat, ~val , initialized in

lines 10-12). In lines 24-27, an infinite routine is defined that assigns values
to the variables by retrieving the value on the bus ~ampBus at a rate ~update
and performing a simple mapping between the domain of the amplitude (estimated linearly by Amplitude, but here converted in dB) and pixel dimensions
and color parameters. Note that theoretically the amplitude range in dB would
be $[-96, 0]$, but $[-60, 0]$ is empirically a more appropriate value for this example.

```
 1  ~ampBus = Bus.control(Server.local) ;
 2  ~audio = Bus.audio(Server.local) ;
 3  a = {
 4      var sig = In.ar(~audio) ;
 5      var amp = Lag.kr(Amplitude.kr(sig)) ;
 6      Out.kr(~ampBus, amp) ;
 7  }.play ;

 9  ~update = 0.1 ; // update rate
10  ~dim = 50 ; // circle radius
11  // color parameters
12  ~hue = 0; ~sat = 0.7; ~val = 0.7;
13  // window and drawing function
14  w = Window("tester", Rect(10, 10, 500, 500))
15  .background_(Color(1,1,1,1)).front ;
16  w.drawFunc = {
17      var oo = 250-(~dim*0.5) ; // to place the circle in the center
18      Pen.addOval(Rect(oo, oo, ~dim, ~dim)) ; // the circle
19      Pen.color_(Color.hsv(~hue, ~sat, ~val)) ; // colors
20      Pen.fill ; // fill it
21  } ;

23  // an infinite routine to update the window
24  {
25  inf.do{
26          // for infinite times look into ~bus
27          ~ampBus.get({ arg amp ; // it represents the value in ~bus
28              // conversion in db
29              ~dim = amp.ampdb.linlin(-60, 0, 10, 500) ;
30              ~hue = amp.ampdb.linlin(-60, 0, 0.4, 0) ;
31              ~sat = amp.ampdb.linlin(-60, 0, 0.8, 1) ;
32              ~val = amp.ampdb.linlin(-60, 0, 0.8, 1) ;
33              {w.refresh}.defer ; // update the window
34          }) ;
35          ~update.wait ;
36      }
37  }.fork ;
```

At this point it is possible to write on the control bus ~ampBus. The first synth x performs a test in which the mouse on the y axis controls the amplitude of a sine wave. By shifting the mouse vertically, the size and color of the circle changes. The synth is deallocated (7), and the variable is assigned to a new synth, reading from a buffer (12), on which a soundfile has been loaded (9). Note that the update rate increases (by reducing ~update, 10), and the synth writes on two audio channels, 0 and the private bus that is routed to the analysis synth a.

```
1  x = {
2      var sig = SinOsc.ar(mul: MouseY.kr(0,1)) ;
3      //Out.ar(0, sig) ; // we don't send it out
4      Out.ar(~audio, sig) ; // but we just write it to bus
5  }.play ;

7  x.free ; // we free the synth, but the routine is still on

9  ~buf = Buffer.read(Server.local, Platform.resourceDir
10     +/+ "sounds/a11wlk01.wav") ;
11 ~update = 0.025 ; // more precise
12 x = {
13     var sig = PlayBuf.ar(1, ~buf, loop:1) ;
14     Out.ar([0, ~audio], sig) ; // on first public bus and on our private bus
15 }.play ;
```

Obviously, the values of the signals "caught" by ~bus.get, and from then on available to the client side –this is the point– do not have to be used for graphics at all. They are available for any required, client-side computation.

9.2 From server to client: use of OSC messages

Until now, it has been stated many times that the client and the server communicate via OSC messages. But so far these messages have remained completely hidden. Where are they? In fact, in our privileged, "linguistic" approach, messages are hidden beneath a layer prepared by the SC language.

The OSC protocol, originally developed for audio applications (the acronym stands for Open Sound Control) is now a standard in multimedia communication between applications belonging to various domains, including audio (SuperCollider, Max/MSP, PD), graphics (Processing), image processing (Eyesweb, VVVV), programming languages (practically all, e.g. Python, Java, Ruby), and also advanced multitrack audio environments (for example, OSC is supported by Ardour and Reaper). It is a protocol that does not provide any semantics, but rather a syntax to create messages. In other words, OSC states *how* messages should be written to be "well-formed", but not *what* they should say. The semantics is defined by the application that sends or receives the message. So, scsynth communicates via OSC, but defines its own semantics: it establishes what certain messages mean in relation to its own ontology as an audio server. This is not a place for a tutorial on OSC: the reader should look on the web for information and tutorials. Meanwhile, to get an idea of the client-server communication, the following code can be evaluated:

```
1  OSCFunc.trace(true) ;
2  // enough
3  OSCFunc.trace(false) ;
```

The class OSCFunc is the general manager of the OSC communication on the client side. In other words, with OSCFunc the user can define what happens in case sclang receives OSC messages. A very useful feature is defined by the method trace, which according to its boolean argument, prints (or not) on the post window all OSC messages received by sclang (and therefore potentially interpretable). If line 1 is evaluated, a situation similar to the following should appear on the post window:

```
 1   OSC Message Received:
 2      time: 111609.33037712
 3      address: a NetAddr(127.0.0.1, 57110)
 4      recvPort: 57120
 5      msg: [ /status.reply, 1, 0, 0, 2, 63, 0.031290706247091, 0.16726991534233,
 6      44100, 44099.998039566 ]

 8   OSC Message Received:
 9      time: 111610.0269853
10      address: a NetAddr(127.0.0.1, 57110)
11      recvPort: 57120
12      msg: [ /status.reply, 1, 0, 0, 2, 63, 0.037191119045019, 0.12907001376152,
13      44100, 44100.000112071 ]
```

Similar messages will follow. These messages are the ones that scsynth regularly sends to sclang to indicate that the communication between the two is active.

Sclang warns that a message has been received: time represents the time since the interpreter has been active. The message is sent from the sender 127.0.0.1, 57110, where the first element indicates an IP address that represents the local communication on the same machine, and the second value is the port from which it has been sent. Remember that when scsynth is booted, it is assigned by default the port 57110 (the one that is printed on the post window while the server boots). The message is received on the port 57120, that is instead assigned by default to the sclang process. This means that if the user wants to send a message to sclang from another application, s/he must use the port 57120 as its port address. The actual message follows (5): it includes a name (/status.reply, note the character / at the beginning of *all* OSC messages by protocol definition) and a set of elements, all packed into an array. In this case, the message is sent by scsynth to tell sclang that it is active, and some basic information about its state.

In the next example, the Ugen SendReply is introduced. It belongs to a family that we never encountered before, that of UGens that deal with sending OSC messages *from* the server. SendReply sends messages back to the client at every trigger: the latter is here specified by the first argument that the chosen trigger UGen, Impulse, receives: in the example, 10 times per second. The second argument indicates the name of the message that is sent, /ping. The third argument is a value, which here is retrieved from the signal (generated on the server, of course) by WhiteNoise. In other words, every tenth of a second, a message /ping that contains the sampled amplitude value of the generated white noise signal is sent back to the client. Note that the SynthDef does not generate audio (better: it calculates an audio signal to send its amplitude to sclang, but also a different value could be sent, e.g a constant numeric value). If the OSC messages are traced by OSCFunc.trace and the synth x is allocated, it becomes possible to see what is going on in the communication. It is time to exploit it, and OSCFunc serves this need. The next example reimplements the one already discussed above.

```
1  (
2  SynthDef(\impl , {arg in = 0, out = 0, dbGain = 0;
3      var sig = LFNoise2.ar(0.5) ;
4      Out.ar(0, SinOsc.ar(sig.linlin(-1, 1, 48, 72).midicps, mul:0.2)) ;
5      SendReply.ar(Impulse.ar(10), '/amp', values: sig)
6  }).add ;
7  )

9  (
10 x = Synth(\impl ) ;

12 i = 0;   v = 0.0  ;
13 w = Window.new("", Rect(10, 10, 500, 500)).front ;
14 u = UserView(w, Rect(0,0, 500, 500)).background_(Color.grey(0.75)) ;
15 u.drawFunc_{
16     Pen.fillColor = Color.red(0.5);
17     Pen.addOval(Rect((i%500), v.linlin(-1.0,1.0, 0,500), 2,2)) ;
18     Pen.fill
19 } ;
20 u.clearOnRefresh_(false) ;
21 o = OSCFunc({ |msg|
22     v = msg[3] ;
23     i = i+1 ;
24     {u.refresh}.defer
25        }, '/amp');
26 )
```

This time the signal sig is used to control SinOsc that is immediately routed on the bus 0. Then, SendReply sends a message /amp back to the client including the value of the signal sig. The only new thing in the rest of the code is OSC-Func. Starting from the last argument, it can be seen how it responds only to messages with the name /amp. If messages with a different name would arrive to sclang, they would simply be ignored. Rather, if a message /amp is received, the function is evaluated. The latter has as its default argument the message itself (here named msg) which thus becomes accessible within the function. In the message, the fourth element is the value of sig, which is assigned to v, and used by drawFunc to plot (as in the example above). Note that by running Cmd +. the OSC responder is removed.

In the next example, communication takes place in two steps, from server to client and from this back again to the server. The SynthDef player simply reads from the buffer (as usual, b) and envelops the signal with a percussive envelope, which deallocates the synth at its end. The SynthDef listener is more interesting. The SynthDef reads from the first input of the sound card (on a standard computer, the microphone) and analyzes the incoming signal through Onsets, a UGen that detects an attack in the input signal (by performing also a FFT analysis, as shown). Onsets acts as trigger for SendReply. In other words, for every retrieved attack, Onsets triggers SendReply to send a message '/attack' to sclang. The message contains the value of Loudness, a UGen (at control rate) again based on FFT like Onsets (and, in fact, it is simply passed chain): Loudness performs an estimate of the "loudness', i.e. the perceived volume, in a range from 0 to 100. On the client side, each time a message '/attack' is recevied, OSCFunc reads the received value (i.e. the computed loudness) and maps it between 1 and 2 (25). This value becomes the control parameter for the argument rate of a synth of type player that is generated (and here the implicit communication will go from sclang to scsynth). Essentially, for every attack that is detected in the microphone signal a "note" is generated, and its pitch varies according to the volume of the input signal. It is a sort of Geiger counter reacting to attack/loudness.

```
1  (
2  b = Buffer.read(s, Platform.resourceDir +/+ "sounds/a11wlk01.wav");

4  SynthDef(\player , { arg buf, out = 0, rate = 1 ;
5      Out.ar(out,
6          FreeVerb.ar(in:PlayBuf.ar(1, buf, rate)
7              *EnvGen.kr(Env.perc, doneAction:2)*2)
8          )
9  }).add ;

11 SynthDef(\listener , {
12     var sig = SoundIn.ar(0) ;
13     var loc = LocalBuf(1024, 1) ;
14     var chain = FFT(loc, sig);
15     SendReply.kr(
16         Onsets.kr(chain, 0.75, \rcomplex ),
17         '/attack',
18         Loudness.kr(chain));
19 }).add ;
20 )

22 (
23 OSCFunc({
24     arg msg;
25     var rate = msg[3].linlin(20, 40, 1, 2) ;
26     Synth(\player ).set(\rate , rate, \buf , b) ;
27 }, '/attack');

29 Synth(\listener )

31 )
```

9.3 OSC to and from other applications

The object OSCFunc is obviously the key to communicate with sclang via OSC. As discussed, sclang usually receives on port 57120. It is therefore possible to send messages to sclang from other applications on a network. For example, let us suppose to have a smartphone and a local network. Obviously, it is necessary to know some data about the local network. In the example, the computer that hosts SC has an address 172.20.10.3 (data recoverable from various system utilities). A popular OSC application for mobile devices is TouchOSC, which allows for the use of a touch interface of smartphones and tablets as an external controller. It is therefore necessary to set in TouchOSC the network address of the computer hosting SuperCollider and the port where to send OSC messages (in our case, 172.20.10.3 and 57120, where sclang receives). By using OSCFunc.trace some useful information can be spotted. For example, the address from which the messages originate (in the specific case, 172.20.10.1). Also of interest is the name of the sent message, as in TouchOSC it is not specified elsewhere: for example '/1/fader1' is associated with a slider in one of the available GUI screens. The following code intercepts these messages and prints their content. It is trivial to change the function to perform other actions (e.g. audio related).

```
1  OSCFunc(
2       { arg msg, time, addr, recvPort;
3            [msg, time, addr, recvPort].postln;
4  }, '/1/fader1') ;
```

But OSC communication does not have to be one-way, on the contrary it can be bidirectional. Sending messages from sclang to other addresses is possible: indeed, it is what normally happens in communication between sclang and scsynth. The next example shows a new object which is tailored toward network communication, NetAddr, which lets us specify an address in terms of ip and port, and send messages to it.

```
1  ~phone = NetAddr.new("172.20.10.1", 9000);

3  ~phone.sendMsg("/1/fader1", 0);
4  ~phone.sendMsg("/1/fader1", 0.5);
```

Here ~phone is the device that was sending OSC messages in the previous example: the variable is assigned an object of type NetAddr that includes the address (the one already discussed) and a port number where to send the message: here we hypothetically use 9000, but the identification will depend on the settings on the TouchOSC side on the device. At this point it becomes possible to send OSC messages to ~phone: the name of the message is the one that represents the graphic slider that in the previous example was sending messages. Obviously, the semantics of the message depends in this case on TouchOSC, which allows a two-way communication by means of the same message: if the slider position is changed, TouchOSC sends a message with the name and value; if TouchOSC instead receives a message with the name and a value, then it changes the position of the slider accordingly to the value. The situation is depicted in Figure **9.1**.

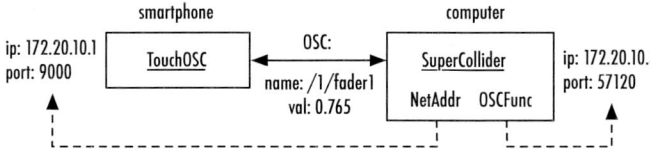

Figure 9.1   OSC communication between a computer and a mobile device.

This example is a bit complicated because its implementation requires a reference to an external object, with the consequent complications (here, TouchOSC). However it made it possible to clarify how network communication can be defined. The next example (from the documentation of NetAddr) works by default because here sclang communicates with itself, but it becomes understandable only in the light of what has been discussed before.

```
1  n = NetAddr("127.0.0.1", 57120); // 57120 is sclang default port
2  r = OSCFunc({ arg msg, time; [time, msg].postln }, '/good/news', n);

4  n.sendMsg("/good/news", "you", "not you");
5  n.sendMsg("/good/news", 1, 1.3, 77);
```

The local 'loop back' address is always 127.0.0.0 or 127.0.0.1. Line 1 specifies via NetAddr to which address sclang will send: that is, the same machine on the port 57120. Then an OSCFunc is defined (2), that responds (on sclang) to messages '/good/ news'. At this point (4, 5), sclang sends to the address specified by NetAddr two messages. But then, two messages will arrive to sclang (as a receiver) on the port 57120, and consequently the function defined in OSCFunc will be executed.

9.4 The MIDI protocol

The most widely used mode for communicating between audio devices is probably still the MIDI protocol, in use since the early 80s.

The MIDI protocol is in fact the standard for communication between music hardware and it is implemented in all dedicated softwares. As with the OSC protocol, this is not the place for a detailed discussion about MIDI. SuperCollider provides a substantially complete implementation of the protocol, both in input and in output, although probably in this second case the implementation is less exhaustive: the most typical use of MIDI in SC is indeed to allow communication from external devices (typically, gestural controllers) to SC. From a hardware perspective, MIDI devices initially required a specific connector, but now they simply use the USB port.

In order to be able to use MIDI, it is first necessary to activate a dedicated software component, MIDIClient: then, it is possible to access the available devices (if any).

```
1  MIDIClient.init ;
2  MIDIIn.connectAll ;
```

The post window typically prints something like the following, which refers to the connected device.

```
1  MIDI Sources:
2     MIDIEndPoint("USB X-Session", "Port 1")
3     MIDIEndPoint("USB X-Session", "Port 2")
4  MIDI Destinations:
5     MIDIEndPoint("USB X-Session", "Port 1")
6  MIDIClient
```

At this point, the SuperCollider approach to MIDI input communication luckily follows the model already seen for OSC: the MIDI counterpart of OS-CFunc is MIDIFunc. Similarly to OSCFunc, MIDIFunc is equipped with a method trace that allows to track the communication in input (and clearly, also the absence of the same). This is a very useful tool because, just as in the case of OSC, there are often unknown features in the messages sent by external MIDI devices. The example below shows the result of trace. Three messages are displayed: they includes the type of the message (control, noteOn, noteOff), the channel on which it is received (this is a feature of MIDI), the number and the value. Typically (but not always) the channel is 0 or homogeneous for the device in use, while the number lets us discriminate the identifier of the message. By definition of the MIDI protocol, the value is in the range $[0, 127]$, i.e. it has a resolution of 7 bits (128 available values, not a large resolution in many cases). As shown, the last two messages have the same number (43) but a value of respectively 127 and 0. The values actually refer to a button pressed on a hardware controller which activates a noteOn message while pressed and a noteOff message on release. Here the values 0 and 127 are just the equivalent of on/off. The number is the same precisely because the two messages are associated on the controller to the same physical object (but this is still a choice, even if a sensed one, in the design of the device). The first message, of control type, instead originates from the rotation of a knob.

```
 1  MIDI Message Received:
 2      type: control
 3      src: -2025843672
 4      chan: 0
 5      num: 26
 6      val: 88

 8  MIDI Message Received:
 9      type: noteOn
10      src: -2025843672
11      chan: 0
12      num: 43
13      val: 127

15  MIDI Message Received:
16      type: noteOff
17      src: -2025843672
18      chan: 0
19      num: 43
20      val: 0
```

As it happened in the case of OSCFunc, with trace incoming MIDI messages can be explored, and then used with the MIDIFunc. This class has a set of methods already arranged according to the most used MIDI types. The MIDI protocol defines messages such as note on, note off, control, touch, bend, polytouch. The following example shows two uses of MIDIFunc.cc (catching MIDI control messages). In both cases, the object prints the received message on the post window. In the first example, it responds to all the incoming control messages, in the second one only to those with number 26 on channel 0 (as in the case of the previous knob).

```
1  MIDIFunc.cc({arg ...args; args.postln});
2  MIDIFunc.cc({arg ...args; args.postln}, 26, 0);
```

The next example discusses the connection of a controller (it is the same hardware controller as above) with the knob associated with the number 26, for the management of audio parameters. The SynthDef midiSynth has two parameters, a frequency for a square wave generator and a rate that controls the panning oscillation between the left and right channels. These two parameters are associated with the knob on the MIDI controller. At each change of position in the knob, a message is sent and received by MIDIFunc, which evaluates the function, and then updates the two parameters by passing the values to the synth x.

```
1  (
2  SynthDef(\midiSynth , {|freq = 440, rate = 1|
3      Out.ar(0,
4              Pan2.ar(Pulse.ar(freq)*LFSaw.ar(rate).unipolar,
5                  LFPulse.kr(rate)
6              )
7          )
8  }).add ;
9  )

11 x = Synth(\midiSynth ) ;

13 MIDIFunc.cc({arg ...args;
14     var v = args[0] ;
15     x.set(\freq , ((v*0.5)+30).midicps,
16         \rate , v.linlin(0,127, 1, 10))
17     }, 26, 0) ;
```

MIDI output communication uses the object MIDIOut, that is implemented at a a lower level (i.e. closer to the hardware). Its use is not particularly complex, as can be seen in this example from the documentation:

```
1  MIDIClient.init;

3  m = MIDIOut(0);
4  m.noteOn(16, 60, 60);
5  m.noteOn(16, 61, 60);
6  m.noteOff(16, 61, 60);
7  m.allNotesOff(16);
```

In order to send messages to a MIDI device, it is necessary to instantiate an object MIDIOut from the available MIDI devices: 0 indicates in fact the first device listed by evaluating MIDIClient.init. It is therefore possible to send messages to the object by using a typical SC interface (i.e., methods such as noteon, noteoff etc). It must be noted that, in the current SC implementation, MIDI output communication is more related to the set of features of the MIDI hardware in use and to the reference operating system. Additionally, MIDIOut is not considered very reliable in terms of accurate timing.

9.5 Reading and writing: File

A classical form of communication in computer science uses reading and writing files. The class SoundFile, specialized for audio files, has already been discussed. Here it is worth introducing a generic, but very useful, class –File– for writing/reading files in ASCII (i.e. "text") and binary format. The following is an excerpt from the hourly weather report for the Turin airport, obtained from an internet query. The first column represents the time of retrieving, the third and fourth, respectively, temperature and humidity.

1	02:50	1014	9	87	VAR-2	Buona	Sereno	-
2	03:50	1014	8	93	VAR-2	Buona	Sereno	-
3	05:20	1014	8	81	-	Buona	Sereno	-
4	05:50	1014	7	81	W-3	Buona	Sereno	-
5	06:20	1014	8	75	WNW-3	Buona	Sereno	-
6	06:50	1015	8	87	VAR-2	Buona	Sereno	-
7	07:20	1015	12	66	-	Buona	Sereno	-
8	07:50	1015	12	71	VAR-2	Buona	Sereno	-
9	08:20	1015	13	66	-	Buona	Sereno	-
10	08:50	1015	14	62	SSE-3	Buona	Sereno	-
11	09:20	1015	14	62	ESE-3	Buona	Sereno	-
12	09:50	1015	15	62	-	Buona	Sereno	-
13	10:20	1014	15	58	E-2	Buona	Sereno	-
14	10:50	1015	16	55	ESE-3	Buona	Sereno	-
15	11:20	1014	16	59	S-4	Buona	Sereno	-
16	11:50	1014	17	51	SE-4	Buona	Sereno	-
17	12:20	1014	18	52	S-5	Buona	Sereno	-
18	12:50	1014	18	55	VAR-2	Buona	Sereno	-
19	13:20	1013	18	55	ESE-4	Buona	Sereno	-
20	13:50	1013	18	52	E-6	Buona	Sereno	-
21	14:20	1013	19	48	E-5	Buona	Sereno	-
22	14:50	1012	20	45	ENE-4	Buona	Poco nuvoloso	-
23	15:20	1012	19	48	E-5	Buona	Poco nuvoloso	-
24	15:50	1012	19	48	ENE-3	Buona	Poco nuvoloso	-
25	16:20	1012	19	45	ENE-3	Buona	Poco nuvoloso	-
26	16:50	1012	18	55	ENE-5	Buona	Poco nuvoloso	-
27	17:20	1012	17	55	SE-7	Buona	Poco nuvoloso	-
28	17:50	1012	15	62	SE-6	Buona	Poco nuvoloso	-
29	18:50	1013	14	67	S-4	Buona	Poco nuvoloso	-
30	19:20	1014	14	67	S-4	Buona	Nubi sparse	-
31	19:50	1013	14	67	SSW-4	Buona	Poco nuvoloso	-
32	20:20	1014	13	71	SSW-6	Buona	Poco nuvoloso	-
33	20:50	1014	13	76	SSW-5	Buona	Poco nuvoloso	-
34	21:20	1015	13	76	SSW-3	Buona	Poco nuvoloso	-
35	21:50	1013	12	81	WSW-4	Buona	Poco nuvoloso	-
36	22:20	1014	12	81	SW-3	Buona	Poco nuvoloso	-
37	22:50	1014	11	87	SSW-3	Buona	Poco nuvoloso	-
38	23:20	1015	11	87	-	Buona	Poco nuvoloso	-
39	23:50	1014	11	87	-	Buona	Nuvoloso	-
40	00:50	1014	11	87	NE-2	Buona	Nuvoloso	-
41	01:20	1014	11	87	-	Buona	Nuvoloso	-
42	01:50	1014	10	93	NE-3	Buona	Nubi sparse	-

The ability to read files in SC allows easy access to various data that can be processed in order to reconstruct a certain data structure. Let us suppose we have stored the previous data in a text file. In the following example, the object File f accesses the text file at the path p in read-only mode, as indicated by "r". The method readAllString returns the entire content of the file as a string, assigned to t (4). Then the file p is closed. The following four lines are dedicated to string-content processing with the aim of reconstructing a data structure (of course, known in advance). Line 6 uses split: invoked on a string, it returns an array of substrings that are split (in this case) when the character $\n is detected, that is the "invisible" character for a line break[1]. In this way, we get an array of strings representing lines from the file. Then, through the method collect each line becomes an array of elements, in this case separated by the tab character ($\t) (7). Rows and columns are now reversed, and will therefore be an array of values of a homogeneous type (8). Finally, only the columns related to temperature and humidity are chosen, converted into integers (since they are, in fact, still strings).

Note that in this way we have gone from ASCII characters in a file to a data structure of integers[2].

```
1  p = "/Users/andrea/SC/introSC/code/comunicazione/dati/tocas01042014" ;

3  f = File(p, "r") ; // open the file
4  t = f.readAllString ; // entire content as a string
5  f.close ; // close the file
6  t = t.split($\n ) ; // we get an array of lines
7  t = t.collect{|i| i.split($\t )} ; // each line -> an array of elements
8  t = t.flop ; // we swap rows and columns
9  t = t[2..3].asInteger ; // we keep  only temperature and humidity
```

The whole can be more elegantly encapsulated into a function, like ~meteo in the following example, which requires a path as an argument and returns the discussed data structure:

---

[1] The character $ indicates the class Char in SC.

[2] Moreover SC defines other classes that make it easier to read from file in a specific format, for example, FileReader, TabFileReader, CSVFileReader.

```
1  ~meteo = {|path|
2      var f = File(path, "r") ;
3      var t = f.readAllString ; f.close ;
4      t.split($\n ).collect{|i| i.split($\t )}.flop[2..3].asInteger ;
5  } ;
```

Finally, the retrieved data can be represented through sound, that is, with a procedure known as "sonification". In sonification, a set of data is displayed through the sound, rather than e.g. graphically. In the following example, the SynthDef exploits an additive approach, in which, in addition to the fundamental frequency, the fact argument determines the rate of damping of the harmonic components, a feature which could be defined as brightness.

The parameter m is then calculated, resulting from ~meteo (9). The next routine sequentially reads couplets of temperature/humidity and associates them to frequency and brightness. In this way, it is possible to grasp by ear that there is a progressive increase in temperature and a decrease in humidity (the sound events grow in pitch and brightness) followed by an opposite movement, precisely in accord to what occurs in the daily data.

```
1   SynthDef(\meteo , { |freq = 440, fact = 0.1|
2       var sig = Mix.fill(40, {|i|
3           SinOsc.ar(freq*(i+1))
4           *(i.neg*fact).dbamp
5       })*EnvGen.kr(Env.perc, doneAction:2) ;
6       Out.ar(0, FreeVerb.ar(sig))
7   }).add ;

9   m = ~meteo.(p) ;

11  {
12      m.flop.do{|i|
13          Synth(\meteo , [
14              \freq , i[0].linlin(-10, 40, 48, 98).midicps,
15              \fact , i[1].linexp(30, 90, 0.1,4)]) ;
16          0.15.wait
17          }
18  }.fork ;
```

Through `File` it is also possible to write data, as shown in the next example, in which a verse from a poem by Toti Scialoja is written on a file. Note that the file extension is irrelevant (and absent here). In the second block the previously written file is read from the disk and its contents assigned to the variable~text.

```
1  (
2  t = "Il coccodrillo artritico che scricchiola" ;

4  p = "/Users/andrea/coccodrillo" ;
5  f = File(p, "w") ; f.write(t); f.close ;
6  )

8  (
9  f = File(p, "r") ;
10 ~text = f.readAllString ; f.close ;
11 )
```

The entire poem by Scialoja is the following:

Il coccodrillo artritico che scricchiola
arranca lungo il greto verso un croco
giallo cromo, lo fiuta, fa una lacrima
se il croco raggrinzisce a poco a poco.

Here we find another example of sonification, this time related to the alphabetical data. The example assumes that the variable ~text is assigned all the text from Scialoja's poem. The (multi-line) string is read and then progressively written onto a textual GUI. Every event-letter is also "played".

```
 1  (
 2  SynthDef(\reader ,
 3  { arg freq = 100, vol = 1;
 4      Out.ar(0,
 5      Pan2.ar(
 6          FreeVerb.ar(
 7              MoogFF.ar(
 8                  Pulse.ar(freq, mul:1)
 9                  *
10                  EnvGen.kr(Env.perc, timeScale:2, doneAction:2),
11                  freq*2,
12          )
13      ),
14      LFNoise1.kr(1),
15      vol
16      )
17  )}).add
18  )

20  (
21  ~time = 60/72/4 ;
22  ~minus = 20 ;

24  w = Window.new("da Toti Scialoja, Il gatto bigotto (1974-1976)"  ,
25      Rect(0,0, 500, 300)).front ;
26  d = TextView.new(w, w.view.bounds)
27      .stringColor_(Color(1,1,1))
28      .background_(Color(0, 0, 0))
29  .font_(Font.monospace(16)) ;

31  // routine scans the text with rate = ~time
32  Routine({
33      var txt = "" ;
34      ~text.do {arg letter, index ;
35      var f = (letter.ascii-~minus).midicps ; // ascii
36      txt = txt++letter ; // text is incremented
37      d    .string_(txt) // it replaces the previous text
38          .stringColor_(Color.hsv(0, 0, 1-(index/~text.size*0.8))) ;
39          // everything gets progressively darker
40      Synth(\reader , [\freq , f.max(20)]) ;
41      ~time.wait ;
42      };
43      1.wait;
44      w.close // the window is closed
45  }).play(AppClock) // we need AppClock for the GUI
46  )
```

Without entering into details, here are just a few considerations: ~time is expressed in sixteenths with bpm mm 72; ~minus is a transposition factor; pitch is obtained by converting each character of the text into its (numerical) ASCII value (34). The Reader is invited to further examine this example.

9.6 Pipe

One of the interesting applications of File is to use SuperCollider as a "scripting" or "gluing" language. A scripting language is one that serves as a high-level control for another language. The user writes in the scripting language, which internally manages the communication with the other language. "Gluing" indicates the use of a language to "glue" together in a unitary form functions or activities performed by other languages or applications. For example, Postscript is a programming language for graphics. A Postscript file contains Postscript code (i.e., simply ASCII text, such as the SuperCollider code) that, once interpreted, returns an image file as its output. If a Postscript file is open with the SC IDE, the Postcript code will show up. The following fragment is an example of a minimal Postscript code. Once copied into a text file and saved with the extension ps, we get a Postscript file. By opening the file (and assuming that a PostScript interpreter is available, as typically happens on most operating systems), its contents are interpreted and the resulting image is generated. In this case the Postscript interpreter will produce from the code the drawing of a line.

```
1  %!
2  144 72 moveto
3  288 216 lineto
4  stroke
5  showpage
```

An example of application of such an approach is this book itself. All the example blocks (both syntax colored code and post window outputs) were generated automatically from the SuperCollider code, through a custom class that analyzes the SC code and generates the appropriate Postscript code. The class

is responsible for introducing in the Poscript file the colors of text and background, the colored frame, and the line numbers. Analogously, in this book all
the figures plotting actual signals were generated from data structures in Super-
Collider, analyzed to produce the required Postscript code. Such an application
is pretty obvious if the reader thinks of the previous Postscript code as a a text
string: with File, that code string can be written onto a file with the extension
ps. This possibility is widely usable for all those parts in a program that are
descriptive, and can be used, for example, to generate code in many environments that are provided with textual interfaces, such as Processing, NodeBox,
R, and so on. At the extreme, it is indeed possible to generate, from within SC,
files containing SC code.

The next example demonstrates how to write a file from SuperCollider
(scripting), and then call a program referring to the same file (gluing).

```
 1  ~psCode = " %! 144 72 moveto 288 216 lineto stroke showpage ";

 3  ~path = "/Users/andrea/minimal.ps" ;

 5  ~file = File(~path, "w") ;
 6  ~file.write(~psCode);
 7  ~file.close ;

 9  ~pipe = Pipe("pstopdf %".format(~path),"w") ;
10  ~pipe.close ;
```

The variable ~psCode is a string that contains the Postscript code discussed
earlier, while ~path is a path on the file system. The block 11-13 opens a file
from ~path, writes the code, and closes it. Lines 15-16 use a new object, Pipe,
that allows to intercept the terminal, an interaction mode for the operating system that, rather than using GUIs, is based on the so-called shell (as it happened
in the past). User interaction with the shell occurs by writing command lines.
Rather than opening a terminal and writing by hand the desired commands,it is
possible to ask SuperCollider to do it for us. Pipe is the object designed to perform such a task: it opens a "pipe" (hence the name) to communicate with the
shell. In the example, Pipe writes on the shell (note the parameter "w", which
stands for "write") the string pstopdf concatenated with the path ~path. The
string is obtained by calling the method format that simply replaces % with the

first passed argument. The string in question, which could obviously be written directly into a terminal, calls the program pstopdf, which converts a PostScript file into a PDF file (of course, assuming that the *pstopdf* program is installed on the operating system and available to the shell –as normally it is in POSIX (Unix-like) systems. Line 16 closes the pipe opened with Pipe which is mandatory. If the process has been successful, then the post window will print 0[3]. A simple access to the shell is also possible by means of the unixCmd method, defined on String. The block 15-16 could then be replaced with the following one:

```
1  "pstopdf %".format(~path).unixCmd ;
```

The next example shows another application of Pipe, starting from an example in the relative help file. Given a folder, the function returns an array with the names of the files that it contains.

```
1  ~sampleList = { arg samplesPath ;
2      var p, 1, labelList = [], fileName ;
3      p = Pipe.new("ls" + samplesPath, "r") ;
4      1 = p.getLine ;
5      while({1.notNil}, {
6          1.postln;
7          if ( 1.contains("."))
8              { labelList = labelList.add(1) } ;
9          1 = p.getLine;
10     }) ;
11     p.close ;
12     labelList
13 } ;

15 ~sampleList.("/Users/andrea/")
```

---

[3] This is one of the numeric codes that Unix returns to the user after each process: it indicates the result of the process itself. The use of the Unix shell is of course far beyond this discussion.

The Unix command ls requires to specify a folder location of the file system, and returns a list of the contained files and folders. The function ~sampleList takes as argument a path (as can be seen in line 15). On line 3, Pipe passes the shell the command ls with the path. Note that the second argument of Pipe this time is "r", which means that we want to read back the results on the shell (in Unix parlance, the so-called stdout), that is, the data concerning file names and folders. The variable l is assigned the string returned by the method getLine, which returns, line by line, the output of ls (4). The cycle while runs out the call to stdout as long as getLine has nothing to return. The conditional checks through the presence of the dot when a file or a folder is retrieved (7-10). This idea is of course a questionable heuristics (the files may not have any extension), but it typically works, for example, if you consider audio files. The resulting array can be used to load audio files, to create GUI with file names, etc.

The use of File and Pipe makes the SuperCollider environment extremely flexible, as the latter can be integrated into the vast ecosystem of Unix programs.

9.7 SerialPort

Finally, one last possibility of communication, which further extends the communication options between SC and other environments, is the serial port. The USB (Universal Serial Bus) port is still the industry standard for communication between digital devices. It is based on serial transmission, which in fact, as we have seen, is well suited to replace the hardware layer for the MIDI protocol, originally based also on a serial logic. As previously discussed with respect to MIDI, the operating system recognizes a MIDI device physically connected through the USB port, and thus treats him accordingly. However, the USB port also allows us to connect other, non MIDI, devices tha are constantly growing in their use, especially microcontrollers and single board computers. Examples of the latter category are Raspberry Pi and UDOO, while the microcontroller *par excellence* is Arduino (in all its versions, although there are many other options). Arduino uses the USB port in order to perform program loading: Arduino programs are developed on the host computer in a dedicated development environment, compiled and then sent to the the board via USB port. However, communication with Arduino can also be done interactively in real-time: in other words, Arduino can exchange data, in input and output, with a software on a host computer. That software can certainly be SuperCollider. There are several ways in which communication with Arduino can take place. For example it is possible to hide the communication at the lowest level in order to provide the user with a more friendly interface. A typical approach is to use libraries that must be loaded on Arduino, side to side with SuperCollider companion classes that refer to the former. In other words, the situation is similar to what happens in the communication from sclang to scysnth in the approach favored in this text: the user interacts by taking advanced of a set of methods to access the features of Arduino defined at the SC language level.

However, what is exchanged between Arduino and the host computer is a set of bytes. It is therefore possible (instructive, useful, and sometimes necessary) to directly manage this low-level communication, through the class SerialPort. By evaluating:

```
1 SerialPort.devices ;
```

a list of devices available to the serial interface is printed on the post window. For example:

```
1 [ /dev/tty.Bluetooth-Incoming-Port, /dev/tty.Bluetooth-Modem,
2 /dev/tty.usbmodem641 ]
```

The list includes all the devices that make use of the serial bus (i.e. the communication channel), including services such as bluetooth. The last element of the list is instead an Arduino microcontroller connected to a USB port. Once the ID for the device is known (which is returned as a string), it becomes possible to access it and open the communication.

```
1 ~ard = SerialPort("/dev/tty.usbmodem641", 115200, crtscts:true) ;
```

The object SerialPort features a set of parameters that depend directly on the hardware specifications of the bus. Apart from the name, the example specifies the data transfer rate and a parameter that deals with the the so-called "flow control" (crtscts). Its value true indicates that, if the amount of data to be transferred is higher than the transfer rate, the same will be placed in a buffer to be progressively emptied when resources are available. At this point, it is possible to read and write values (i.e. bytes) through the serial port. The semantics of the communication of course depends on the program loaded on the Arduino side. Let us consider the following example:

```
1  ~port = 2; ~value = 255 ;
2  ~ard.putAll([253, port, value, 255]) ;
```

The code assumes that the program on Arduino accepts 4-byte blocks. The first and the last byte are control bytes: in other words, the presence of two bytes equal to 253 and 255 respectively at the beginning and the end of a block of four bytes indicates to the Arduino program that a message is coming. The other two values, represented by the variables ~port and ~value, are interpreted as values which select an output port on Arduino and set it to a value (expressed with a 8-bit resolution, that is, in the range $[0, 255]$). In other words, the program on the Arduino receives four bytes: if the first and the fourth are respectively 253 and 255, then a port is selected (here, 2) and set with a value (here, 255). Arduino will generate on port 2 (associated with circuits capable of generating PWM signals) a PWM electrical signal of the maximum available quantity. The interpretation of the message sent by SC through the serial port depends on the Arduino code: the latter here has been described only with respect to its logic, otherwise we should have dove into Arduino's language. Note that luckily SerialPort converts integers into bytes. Reversely, it is possible to communicate from Arduino to SuperCollider. An example would be sending to SC the values generated by a sensor connected to Arduino. In that case, SC would read the bytes sent from Arduino and will define how to interpret them. Thus, the communication through the serial port is simple in itself, assuming, however, that the user knows how to write/read the information in the Arduino program.

Finally, the serial port must be closed when the device is to be disconnected. In the example below the first line closes the communication with the previous device, while the second one closes all the ports.

```
1  ~ard.close ;
2  SerialPort.closeAll ;
```

9.8 Conclusions

Classes and objects discussed in this chapter allow us to greatly expand the information ecosystem where SuperCollider is located and operates. There are two important aspects of this expansion. On the one hand, many additional features can be integrated in complex multimedia projects. On the other, symmetrically, SuperCollider's high-level algorithmic approach can be extended to other hardware and software environments.